CONTENTS

KW-266-316

SKETCH MAP OF THE ISLAND OF LUNDY

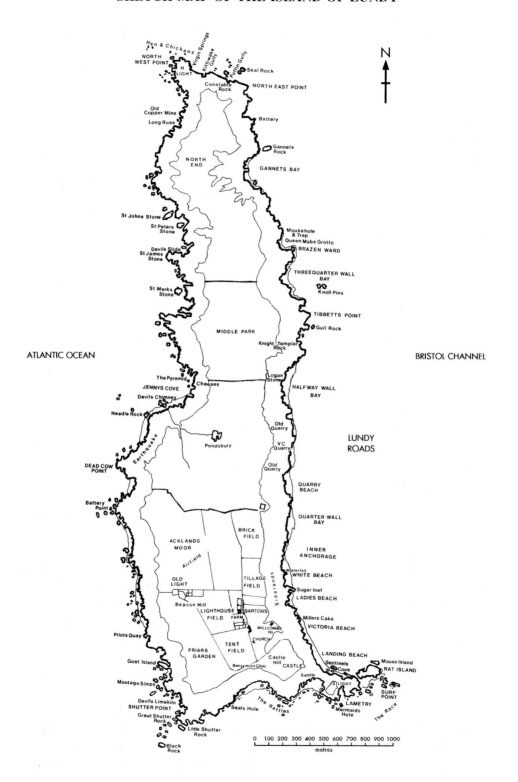

LUNDY STUDIES

CELEBRATING 60 YEARS OF THE LUNDY FIELD SOCIETY

PROCEEDINGS
OF THE 60TH ANNIVERSARY SYMPOSIUM
OF THE LUNDY FIELD SOCIETY

23 September 2006

Ur.

Su!

I

Edited by
Jennifer George

LUNDY FIELD SOCIETY

PROCEEDINGS
OF THE 60TH ANNIVERSARY SYMPOSIUM
OF THE LUNDY FIELD SOCIETY

Dedicated with grateful thanks to
Sir Jack Hayward OBE
President of the Lundy Field Society

First published 2007

© The Lundy Field Society and the contributors

ISBN 978-0-9530532-1-6

Produced by Malago Press and Print Services

FOREWORD

Lundy is a small island 5 km (3 miles) long and 0.8 km (½ mile) wide situated in the approaches to the Bristol Channel, some 18 km (11 miles) from the nearest North Devon mainland. Its sheer granite cliffs rise from sea level to a fairly level plateau where there is archaeological evidence of habitations from the Bronze Age onwards. A castle dating from 1244, sites of medieval farmsteads with associated field systems, and Victorian granite quarries contribute to the interesting facets of the island landscape.

A diverse range of plants and animals occur on the island and in its surrounding waters. Many seabirds breed here and the island is an important stopping point in transit for large numbers of migratory birds. 70% of the island is classified as a Site of Special Scientific Interest (SSSI), and the surrounding seas are England's only Statutory Marine Nature Reserve (1986), containing two scheduled wreck sites. Part of the reserve became a No-Take Zone in 2003 and the reserve was designated an Area of Special Conservation in 2005. Feral goats, sika deer, Soay sheep and the Lundy ponies roam the island, and a unique plant, the endemic Lundy cabbage, *Coincya wrightii*, flourishes on the sheltered east side.

Much of the scientific investigation and research carried out on Lundy has been initiated and monitored by the Lundy Field Society (LFS), which was founded in 1946. In 1996, their 50th anniversary was celebrated by the publication of *Island Studies - Fifty Years of the Lundy Field Society* that featured reports of work on the island's archaeology, geology, history, terrestrial flora and fauna, birds and freshwater and marine life.

In September 2006, to mark the Lundy Field Society's 60th anniversary, a symposium was held at the University of Exeter which was attended by 112 delegates. The papers given and the poster displays presented up-to-date accounts of more recent studies of the marine, freshwater and terrestrial ecology, as well as the history and archaeology of the island.

The LFS Chairman, Roger Chapple, introduced the Symposium and thanks are also due to the chairmen of the three sessions, Henrietta Quinnell (Archaeology and History), Keith Hiscock (Marine and Freshwater Ecology) and Tony Parsons (Terrestrial Ecology). Papers were given by Shirley Blaylock, Myrtle Ternstrom, Robert Irving, Jennifer George, John Hedger and Stephen Compton. The six excellent poster presentations were by Kate Cole; Richard Castle; David Appleton, Helen Booker, David Bullock, Lucy Cordrey & Ben Sampson; David Bullock & Lucy Cordrey; Peter James, Ann Allen & Barbara Hilton; Miles Hoskin, Chris Davis, Ross Coleman & Keith Hiscock.

The symposium was organised by a sub-committee of the LFS led by Myrtle Ternstrom, aided by André Coutanche (organisation of PowerPoint presentations, Symposium pack CD and all things technical), Alan Rowland (Symposium registration), Tony Cutler (venue and speakers), Kate Cole (poster presentations) and Jennifer George (Symposium volume). The following members contributed valuable help on the day: Frances Castle, Roger Chapple, Kate Cole, Marie Jo

Coutanche, André Coutanche, Tony Cutler, Simon Dell, Paul James, David Molyneux, Alan Rowland, Sandra Rowland, Frances Stuart, Myrtle Ternstrom, and Michael Williams.

We extend special thanks to Sir Jack Hayward OBE, President of the Lundy Field Society, for a generous grant that made the Symposium possible, and to whom this volume is dedicated.

Jennifer George
May 2007

INTRODUCTION TO THE SYMPOSIUM

by

ROGER CHAPPLE

Chairman of the Lundy Field Society

Hescott Haven, Hescott Farm, Hartland, Devon, EX39 6AN

e-mail: chapplerj@btinternet.com

Distinguished Guests, Ladies and Gentlemen - Welcome.

The Devon Bird Watching and Preservation Society had frequently used Lundy since their reorganisation in 1929 and in 1945 at the resumption of bird monitoring after the War, they considered expansion by linking their activities in Devon to Lundy and to Scilly and the Pembrokeshire Islands.

With this in mind the Devon Society contacted the owner of Lundy, Martin Coles Harman, who was a keen naturalist and he responded with enthusiasm suggesting to Professor Harvey at the University of Exeter the formation of the Lundy Field Society. He provided the Old Light on Lundy for the use of the Society (free of charge) and gave a cheque for £50. Harman was a man of strong principles and held the independence of Lundy dear to his heart. A Lundy Society would therefore be acceptable whereas a Devon Group simply would not do.

The early establishment by the Lundy Field Society of the Heligoland Traps on the island provided then, as they still do, vital information on the migration of birds. Bird watching and field studies have expanded into wider aspects of field studies, including the marine environment and archaeology.

Members of the Society have been instrumental in setting up the country's first Marine Nature Reserve and others, by way of their earlier archaeological surveys and excavations, have led to the island-wide archaeological survey carried out by the National Trust.

The programme of our meeting today will cover some of these areas. Regrettably, but inevitably, we have had to be selective in the choice of papers and consequently we have concentrated on the more recent work. Some other aspects of the Society's activities are presented in the Poster Displays.

We are now 60 years old and today's Symposium has been arranged as part of our celebrations. We must record our gratitude for the generous support for this meeting given by our President, Sir Jack Hayward.

We look forward to the future eagerly and appreciate the continuing help received from the Landmark Trust, the Lundy General Manager, Derek Green, and the Islanders, and for the faithful support given by our members who regularly attend our conservation groups on the island. We value our position with the Lundy Management Committee and offer our continuing co-operation.

I trust that you will all enjoy our meeting today and that you will leave with an enhanced knowledge of the island and perhaps more importantly a broader knowledge of the activities of our Society.

The Old Light, with Beacon Hill Cemetery in the foreground. The Old
Light was for many years the headquarters of the Lundy Field Society.
Drawing by the late John Dyke

MILESTONES IN THE ARCHAEOLOGY OF LUNDY

by

HENRIETTA QUINNELL

9 Thornton Hill, Exeter, Devon, EX4 4NN

e-mail: H.Quinnell@exeter.ac.uk

ABSTRACT

The paper provides a brief résumé of what is known about the chronology of settlement on Lundy from studies of flint work and ceramics from the end of the last Ice Age until the Early Medieval period.

Keywords: *Mesolithic, Neolithic, Bronze Age, Roman, pottery, lithics*

INTRODUCTION

Over the last sixty years there have been major advances in archaeological techniques and theoretical approaches as well as an exponential increase in the amount of data available world wide. In the same period there has been a range of programmes of archaeological research on Lundy, all sponsored in various ways by the Lundy Field Society, and which reflect these advances. This paper attempts a brief statement on what is reliably known about archaeological chronology on Lundy from the end of the last Ice Age until the Early Medieval period. The 'Milestones' of the title refer to established points in this chronology. There are as yet no radiocarbon determinations from archaeological sites on the island and chronology is applied by analogy from that established across south west Britain.

MESOLITHIC AND NEOLITHIC

Evidence for these periods is restricted to collections of worked flints made over the years, some as organised programmes of field walking and test pit excavation (e.g. Schofield & Webster, 1989; 1990; 1991), others as the recording of chance finds. These lithic finds are currently being studied by Ann and Martin Plummer for the National Trust and their results referenced below as A. & M. Plummer. Almost all the flint used appears to be sourced from beaches in North Devon, probably from Lundy itself. So far no material characteristic of the Early Mesolithic has been found, from the ninth to the mid-eighth millennia B.C.: in general across south west Britain Early Mesolithic flints are few compared to those of the Later Mesolithic (Roberts, 1999). However two pieces may indicate activity during the Late Palaeolithic or the Palaeolithic/Mesolithic transition. The Later Mesolithic is of much longer duration than the Early Mesolithic, from the mid-eighth to the late fifth millennia B.C., and produces much more material and some distinctive microlithic forms. Over most of Lundy, flints tend to be found in broad scatters and not, so far, the tight concentrations which may indicate the positions of house sites. Current studies (A. & M. Plummer) indicate activity in the southern and northern areas of the Island with less evidence in the central area between the Quarter Wall and Threequarter Wall. There was extensive use of the Tillage and Brick Field areas (Schofield, 1992; 1994). The hunter-gather lifestyle of Mesolithic communities involved considerable mobility and Lundy is likely to have been visited on a seasonal basis.

The techniques used to work flint change a little in the Neolithic and this, as well as the introduction of new artifact types, notably single piece arrowheads in place of composite microlithic points, makes it possible to distinguish Neolithic flint artifacts from those of the preceding period. A few pieces of nodular flint, from non-beach sources in Devon, should belong to this period (A. & M. Plummer). The Neolithic covers the fourth and third millennia B.C. A sparse scattering of Neolithic flint across Lundy has been recognized, including a few Early Neolithic leaf arrowheads from the Brick Field and North End, and also a Late Neolithic transverse arrowhead from North End (A. & M. Plummer). None of the distinctive Neolithic ground stone axes has yet been found on the Island; however a fragment of a polished flint axe has been found at North End. There is no pottery apart from two possible sherds from a surface collection at SS13254789 on the North End, found together with a leaf arrowhead (Quinnell, in preparation). Current thinking (Thomas, 1999) sees a substantive element of gathering and mobility in the lifestyle of Neolithic communities and it is reasonable to see the Lundy finds as evidence of some continuity in seasonal resource exploitation from the Mesolithic.

EARLY AND MIDDLE BRONZE AGES

From late in the 3rd millennium B.C., and loosely associated with the introduction of Beaker pottery, round barrows and cairns began to be constructed across south west Britain, most in the first half of the second millennium B.C., the Early Bronze Age, but a few on Exmoor in the second half of this millennium, the Middle Bronze Age (Quinnell, 1997, 34-5). At least fifteen of these sites are recorded in the National Trust Sites and Monuments Record but none have been investigated (but see below Middle Park I). By the Middle Bronze Age evidence of settled farming is found across south west Britain, leaving traces of fields, enclosures and hut circles or platforms marking the landscape. On Lundy systematic field survey by the National Trust 1989-99 has provided a record of all archaeological sites marking its surface, from Early Bronze Age cairns and barrows, through the complexities of settled farming from the Middle Bronze Age onwards to the fields, settlements and industrial activities of Medieval and later period (Thackray, 1999: Blaylock, this volume).

A number of hut circles were excavated by K.S. Gardner in the 1960s (Gardner, 1965; 1967; 1969) but have remained unpublished. Recent work by the author (Quinnell, in preparation) has enabled the ceramics from these excavations to be identified and provided with a broad chronology. The earliest of this material is the group of Biconical domestic ware from North End Hut Circle 6 (Gardner, 1969): other Biconical sherds come from other excavated North End hut circles, as surface finds close to Widow's Tenement and from Test Pit 235 dug by Schofield and

Webster (1991) south of Quarter Wall. This Biconical domestic ware belongs to the Middle Bronze Age with a good comparable assemblage at Brean Down (Woodward, 1990). While some problems about the components of the pottery remain to be resolved by detailed petrological work, both this Biconical domestic ware and the subsequent prehistoric ceramics referred to below appear to have been manufactured on Lundy.

Flint continued to provide a range of tools of types distinctive to the Early and Middle Bronze Ages. Schofield (1992, 71-6) has highlighted a range of this material found across the North End and made accessible by extensive surface burning in 1933 and 1935. The current work by A. & M. Plummer also highlights activity across the North End.

LATE BRONZE AND IRON AGES

Several excavated hut circles have produced Late Bronze Age Plain Ware, a simple ceramic style dating from the eleventh to the ninth centuries B.C.: again a good comparable assemblage with associated radiocarbon dates comes from Brean Down (Woodward, 1990). The hut circles which produced this are those excavated on Beacon Hill by Gardner in 1966 (Gardner, 1967), the nearby hut circle which underlies the Beacon Hill cemetery (Thomas, 1992), hut circles at Middle Park I (Gardner, 1965) and II (Gardner, 1967), and one at the North End. (The character of the site at Middle Park I is unclear: it may possibly be a cairn with subsequent use as a domestic site). This Plain Ware represents the most widespread prehistoric activity so far identified through excavation on Lundy. Both of the Beacon Hill hut circles contain fragments of briquetage in a locally sourced fabric indicating salt production on the Island at this date. Some simple flintwork was still in use in this period.

To date no pottery has been reliably identified as belonging to the Early Iron Age. (It should be noted that the Biconical and Plain Ware groups were initially assigned to the Early Iron Age (Gardner, 1965; 1967; 1969; 1972) as, at the time of their excavation, the detailed prehistoric ceramic sequence in south west Britain was not well understood). In general across Devon and Cornwall little pottery was in use at this period, broadly the eighth to fourth centuries B.C. (e.g. Quinnell, 1999). The Middle Iron Age, from the third to the first centuries B.C. is represented by South Western Decorated Ware (Glastonbury Ware) of which a single sherd came from a North End hut circle.

THE ROMAN PERIOD

Charles Thomas's excavations in 1969 below the Beacon Hill cemetery demonstrated that beneath this, and partly re-using a hut circle with Late Bronze Age Plain Ware, was a circular house or hut circle occupied in the later Roman centuries. The pottery consisted of black-burnished ware from the Poole Harbour area of Dorset, South Devon Ware, probably made in the Dart Valley, and Exeter Gritty Grey Ware, produced somewhere in the Exeter area (Quinnell, 2006), and is likely to range in date from the later second century until the fourth century A.D. *Pace* Thomas

(1992, 45) there are no ceramics made locally, apart from some briquetage indicating salt production, a situation similar to much of Exmoor where pottery was imported from the south coast of Devon and Dorset (pers. comm. L. Bray). The only other distinctive find was a rotary quern. Elsewhere a single sherd of black-burnished ware was recorded from Test Pit 12 just south of Quarter Wall (Schofield & Webster, 1989, 36) and Gardner (1961a) records two grey ware rim sherds without other provenance in Bristol City Museum.

THE POST-ROMAN CENTURIES

The principal evidence is the Beacon Hill cemetery with its four inscribed Christian memorial stones. Thomas (1992, 43) suggests that OPTIMI and RESTEUTA are likely to date to around A.D. 500 on epigraphic style, that POTITI falls in the later sixth century and the fourth, ... IGERNI (FILI) TIGERNI, to the first half of the seventh century. Thomas's excavations demonstrated a complex series of long cist grave burials within an enclosure which he interpreted as belonging to a monastic settlement. No domestic structures associated with the suggested monastery or any others of contemporary date have yet been found. By this stage no pottery was being manufactured in Devon but a very small amount was imported from the Mediterranean. Two imported sherds have recently been identified from Pigs Paradise, effectively unstratified: one is from a Late Roman 1 amphora (British Bii), the other from a buff-coloured amphora (McBride, 2005): the most likely date for these imports is late fifth to early sixth centuries A.D. A third imported sherd, a surface find at SS138444 (Gardner, 1961b: Langham & Langham, 1970, 140) is grouped by Thomas (1981, 25) as a minor unidentified ware. The virtual absence of ceramics in North Devon continues until the occurrence of chert-tempered wares in the eleventh and twelfth centuries but none of these has been found on Lundy so far.

CONCLUSION

It is not suggested that the datable artefacts so far available for study represent the full sequence of past activity on Lundy. It is unlikely that the Island was unoccupied during much of the Iron Age or the early Roman centuries. The framework presented above is only the start of understanding and most of this is based on data which is not yet fully analysed and published. Full publication is the next step in the understanding of this period of Lundy's archaeology. There has been no excavation to modern standards supported by the full repertoire of methods now available for investigating environmental, chronometric and ecofactual data. In time, when adequate resources can be made available, a programme investigating a cairn or barrow and a series of hut circle sites and their related enclosures and field systems will form an invaluable second step to advancing understanding of the settlement sequence on the Island.

ACKNOWLEDGMENTS

I am grateful to Shirley Blaylock and to Ann and Martin Plummer for information and advice towards the completion of this paper.

REFERENCES

Allan, J. & Blaylock, S. 2005. Medieval Pottery and Other Finds from Pigs Paradise, Lundy. *Proceedings of the Devon Archaeological Society,* 63, 65-92.

Gardner, K.S. 1961a. Archaeological note (material in Bristol City Museum). *Annual Report of the Lundy Field Society 1959/60,* 13, 63-4.

Gardner, K.S. 1961b. Dark Age remains on Lundy. *Annual Report of the Lundy Field Society 1959/60,* 13, 53-62.

Gardner, K.S. 1965. Archaeological investigations, Lundy, 1964. *Annual Report of the Lundy Field Society 1963/64,* 16, 30-2.

Gardner, K.S. 1967. Archaeological investigations, 1966/7. *Annual Report of the Lundy Field Society 1965/66,* 17, 31-3.

Gardner, K.S. 1969. Lundy archaeological investigations 1967. *Annual Report of the Lundy Field Society 1968,* 19, 41-4.

Gardner, K.S. 1972. *Lundy: an Archaeological Guide.* The Landmark Trust.

Langham, A. & Langham, M. 1970. *Lundy.* Newton Abbot: David & Charles.

McBride, R.M. 2005 Appendix 3: the Late Roman Amphora sherds. In Allan, J. & Blaylock, S. 2005, 88.

Quinnell, H. 1997. Excavation of an Exmoor Barrow and Ring Cairn. *Proceedings of the Devon Archaeological Society,* 55, 1-38.

Quinnell, H. 1999. Pottery. In T.H. Gent & H. Quinnell. Excavations of a Causewayed Enclosure and Hillfort on Raddon Hill, Stockleigh Pomeroy. *Proceedings of the Devon Archaeological Society,* 57, 38-53.

Quinnell, H. 2006. Prehistoric and Roman Pottery from Lundy. *Annual Report of the Lundy Field Society 2004 ,* 54, 89-92.

Quinnell, H. in preparation. Prehistoric Pottery from Lundy Island. For the National Trust.

Roberts, A. 1999. Late Upper Palaeolithic and Mesolithic Hunting-Gathering Communities. In R. Kain & W. Ravenhill, (Eds). *Historical Atlas of South-West England,* 47-50. University of Exeter.

Schofield, A.J., 1992. The Langham Collection and Associated Finds: a Large Assemblage of Chipped Stone Artefacts from Lundy. *Annual Report of the Lundy Field Society 1991,* 42, 70-84.

Schofield, A.J., 1994. Lithic artifacts from test-pit excavations on Lundy: evidence for Mesolithic and Bronze Age Occupation. Proceedings of the Prehistoric Society, 60, 423-31.

Schofield, A.J. & Webster, C.J. 1989. Archaeological Fieldwork 1988: the Results of Test-pit Excavations and Geophysical Prospection South of Quarter Wall. *Annual Report of the Lundy Field Society 1988,* 39, 31-45.

Schofield, A.J. & Webster, C.J. 1990. Archaeological Fieldwork 1989. Further Test-pit Excavations South of Quarter Wall. *Annual Report of the Lundy Field Society 1989,* 40, 34-47.

Schofield, A.J. & Webster, C.J. 1991. Archaeological Fieldwork 1990: Further Investigations of Artefact Concentrations South of Quarter Wall. *Annual Report of the Lundy Field Society 1990,* 41, 34-52.

Thackray, C. 1999. Completion of the National Trust Landscape History Survey of Lundy, 1989-1999. *Annual Report of the Lundy Field Society 1998,* 49, 48-55.

Thomas, C. 1981. *A Provisional List of Imported Pottery in Post-Roman Western Britain and Ireland.* Redruth: Institute of Cornish Studies Specialist Report 7.

Thomas, C. 1992. Beacon Hill Re-visited: A Reassessment of the 1969 Excavations. *Annual Report of the Lundy Field Society 1991,* 42, 43-54.

Thomas, J. 1999. *Rethinking the Neolithic.* London: Routledge.

Woodward, A. 1990. Bronze Age Pottery. In M. Bell, *Brean Down Excavations 1983-1987,* 121-145. London: English Heritage Archaeological Report 15.

PATTERNS OF SETTLEMENT ON LUNDY:
PUTTING LUNDY'S ARCHAEOLOGY ON THE MAP

by

SHIRLEY BLAYLOCK

The National Trust, Devon and Cornwall Region, Devon Office,
Killerton House, Broadclyst, Exeter, Devon, EX5 3LE
e-mail: shirley.blaylock@nationaltrust.org.uk

ABSTRACT

The landscape features of Lundy show settlement evidence that can be traced from at least the Early Bronze Age (*c*.2000 B.C.) to the present. This paper looks at elements of the National Trust survey of these extensive features, with emphasis on field systems, prehistoric and medieval settlement.

Keywords: *Landscape survey, field systems, Bronze Age, Medieval, settlement*

INTRODUCTION

Lundy has excited archaeological interest for at least 150 years and our understanding of the archaeological remains on Lundy and what they can tell us about past life on the island has developed considerably from what was known 60 years ago when the Lundy Field Society was established. Ten years ago, for the 50th Anniversary of the Society, Caroline Thackray summarised past archaeological work and the nature of the recently completed landscape survey undertaken by The National Trust (Thackray, 1997). At that time we were still very much grappling with the huge volume of detailed data collected from the survey and trying to make sense of it all. All the survey data is now on a database and some interpretations have been worked through, although a full synthesis of the results integrated with past work has not been prepared. Numerous questions have also been raised and some remain unanswered. It has not been possible to publish a full account of the survey and in this paper I will attempt to summarise some of our main conclusions from the survey, combined with results from recent work on excavated material.

The purpose of the landscape survey was twofold. To return to the title of this paper, the main objective was to map the archaeological remains (Figure 1) to know, in the most basic terms, where they were both in relation to each other and the topography. There was a real need to have this map and a written, drawn and photographic record of known or visible archaeological features so that they could be properly managed. In certain cases individual site plans existed and there were spot locations for others but the location of many sites was still problematic and their site plan unrecorded. Parts of the relict field systems had also been mapped but there were many gaps. The second objective was really a by-product of the survey. We wanted to improve our understanding of the landscape and hoped to provide material for updated interpretation by looking at the patterns of settlement and occupation and attempting to understand what this could tell us about the extent and nature of man's activities on the island. It is intended that the survey is a base for future work including additional observations or discoveries. The survey was undertaken as a training exercise for National Trust staff and volunteers, including some members of the Lundy Field Society.

The landscape survey is available on a map- and site-based database held by The National Trust and in the island office and is used for information when works are proposed or to improve management of the archaeological sites, for example by vegetation control (largely bracken and rhododendron) or consolidation of fabric. The survey data was also used as the basis for a new general field guide (The National Trust 2002) and leaflet (The National Trust 2000) and a new interpretation room in the Rocket Shed. Even since these were completed, our understanding of the archaeology has developed perhaps most particularly by the re-examination of the prehistoric pottery by Henrietta Quinnell in the light of increased knowledge since the excavations 40 years ago. A recent 'watching brief' of service trenches has produced a volume of medieval pottery sufficient for detailed scientific analysis not previously undertaken, and this tells us more about the status of the settlement, trade links and markets and brings Lundy into the regional study of medieval pottery currently being developed.

As seems always to be the case with the study of archaeological remains the more we learn the more we realise we do not know. Interpreting individual landscape features is full of difficulties. The interpretation of an earthwork will be developed from its apparent relationship to other landscape features and a comparison with other sites. However later activities such as ploughing, robbing for stone and excavation can change the morphology of the site or feature and make interpretation less secure. Sometimes the interpretation of the function of a feature or site can be satisfactorily determined but the dating of the feature is much more tentative; this is very much the case with parts of the field systems, some of which may have evolved over a long time or been utilised in more than one period. It is also true for the remains of some recorded structures. We can do our best to interpret what we have from our present knowledge, derived from the decades of past work, but also be open to new ideas and research. There is still much to be discovered.

THE EARLIEST PREHISTORIC LANDSCAPE FEATURES

Although material remains from the later Neolithic and Early Bronze Age are scant, Bronze Age burial sites are recorded across the whole island and a small number of standing stones is found south of Quarter Wall. These are site types generally thought to belong to a slightly earlier period than most of the evidence for Bronze Age settlement so far identified. This situation is not uncommon on the mainland and the usual explanation of this may apply here: that the island was used seasonally or had a particular ritual or religious significance for its tribal area, with burial or ritual sites established before a larger, more settled, farming community inhabited the island. A small number of the burial sites may be considered to be prestige cairns

Management of Lundy's Archaeological Monuments

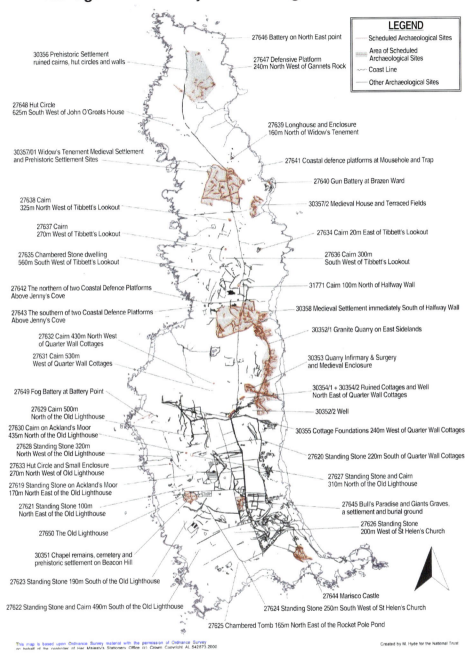

27646 Battery on North East point

27647 Defensive Platform
240m North West of Gannets Rock

30356 Prehistoric Settlement
ruined cairns, hut circles and walls

27648 Hut Circle
625m South West of John O'Groats House

27639 Longhouse and Enclosure
160m North of Widow's Tenement

30357/01 Widow's Tenement Medieval Settlement
and Prehistoric Settlement Sites

27641 Coastal defence platforms at Mousehole and Trap

27640 Gun Battery at Brazen Ward

27638 Cairn
325m North West of Tibbett's Lookout

30357/2 Medieval House and Terraced Fields

27637 Cairn
270m West of Tibbett's Lookout

27634 Cairn 20m East of Tibbett's Lookout

27635 Chambered Stone dwelling
560m South West of Tibbett's Lookout

27636 Cairn 300m
South West of Tibbett's Lookout

27642 The northern of two Coastal Defence Platforms
Above Jenny's Cove

31771 Cairn 100m North of Halfway Wall

30358 Medieval Settlement immediately South of Halfway Wall

27643 The southern of two Coastal Defence Platforms
Above Jenny's Cove

30352/1 Granite Quarry on East Sidelands

27632 Cairn 430m North West
of Quarter Wall Cottages

27631 Cairn 530m
West of Quarter Wall Cottages

30353 Quarry Infirmary & Surgery
and Medieval Enclosure

27649 Fog Battery at Battery Point

30354/1 + 30354/2 Ruined Cottages and Well
North East of Quarter Wall Cottages

27629 Cairn 500m
North of the Old Lighthouse

30352/2 Well

27630 Cairn on Ackland's Moor
435m North of the Old Lighthouse

30355 Cottage Foundations 240m West of Quarter Wall Cottages

27628 Standing Stone 320m
North West of the Old Lighthouse

27620 Standing Stone 220m South of Quarter Wall Cottages

27633 Hut Circle and Small Enclosure
270m North West of Old Lighthouse

27627 Standing Stone and Cairn
310m North of the Old Lighthouse

27619 Standing Stone on Ackland's Moor
170m North East of the Old Lighthouse

27645 Bull's Paradise and Giants Graves,
a settlement and burial ground

27621 Standing Stone 100m
North East of the Old Lighthouse

27626 Standing Stone
200m West of St Helen's Church

27650 The Old Lighthouse

30351 Chapel remains, cemetery and
prehistoric settlement on Beacon Hill

27623 Standing Stone 190m South of the Old Lighthouse

27644 Marisco Castle

27622 Standing Stone and Cairn 490m South of the Old Lighthouse

27624 Standing Stone 250m South West of St Helen's Church

27625 Chambered Tomb 165m North East of the Rocket Pole Pond

Figure 1: A simplified plan of island archaeology derived from the National
Trust measured survey

or major landscape markers as they appear to be deliberately sited in a prominent position. These include the remains of a cairn under John O'Groat's House at the extreme North End; a similar cairn built on a rocky outcrop just south of Threequarter Wall (at times interpreted as a round tower or windmill base, Figure 2); two burial sites on Tibbetts Hill including a cist; a probable cist burial at the south end of the island (now heavily mutilated) just outside the modern field wall enclosure; and possibly in Ackland's Moor. There is the tantalising possibility that a similar cairn or burial was sited on Beacon Hill although no evidence for this has so far been identified. It may be that some of these cairns were intended to be visible from the sea, one only needs to observe the prominence of the Admiralty Lookout on approaching the island to realise this. However, they are also prominent from the land and perhaps would have been much more so when the landscape was devoid of more modern features such as walls and buildings that now attract the eye. Standing on the high points themselves a number of these sites are intervisible. Other less prominent sites have also been recorded: a mound in Widow's Tenement; two burials in Middle Park, one of which is within a kerb or enclosing stone setting; five mounds south of Pondsbury; and possibly a small number of others at the North End.

An enigmatic oval enclosure of individual stones lies just to the north of the water course emanating from Pondsbury. It appears to be prehistoric in type - most likely Bronze Age - but its function is still puzzling. It may be a compound or designated area of either a religious or practical function although the visible remains are too fragmentary to suggest a stock proof enclosure unless it was reinforced with banks or fencing of which there is now no trace.

THE EARLY DEVELOPMENT OF THE FIELD SYSTEMS AND BRONZE AGE SETTLEMENT

If we take away the modern field walls and enclosures, there are strong suggestions that many of the remaining field systems recorded by the landscape survey originated in the Bronze Age. As outlined by Henrietta Quinnell (this volume) a re-examination of the pottery excavated in the 1960s tells us that much of the settlement previously believed to be of Late Bronze Age or Iron Age date is of the Middle or Late Bronze Age contemporary with similar settlements and farming communities on the uplands of the South West. This has led us to conclude that much of the identifiable relict field systems visible across the island could have originated at this time too; many are clearly associated with hut circles or other Bronze Age remains.

The best-known settlement lies at the North End and is of Middle Bronze Age date (Figure 3 and Quinnell, this volume), although interpretation of the individual features within it is not always easy. As indicated by Gardner (1972), the North End appears to be cut off by the location of a wall across the plateau neck at Gannets Combe. Whether this is for stock control, or demarcation for some other form of land division, for example ownership, is not clear. All apart from one identified structure lie on or north of this boundary, perhaps representing five units, sometimes, as described by Gardner, a round structure attached to an apparently rectangular one. One hut circle lies on this wall, four other units are associated with fragmentary

Figure 2: A large cairn south of Threequarter Wall

walls, suggesting one or two enclosures. One of the houses is strangely isolated in a very exposed position at the North End but is associated with domestic pottery (Gardner, Hut 6). The remaining isolated structure at the North End is sub-rectangular and appears to contain a dividing wall. It is undated, leaving the possibility that it could be later, and lies some 280 metres to the south of this boundary. The layout and relationships of these features suggest that the hut circles were built first with the boundary walls being secondary.

The only other location where a group of dwellings forming a small settlement is recorded is on Beacon Hill. Two structures identified as houses have been excavated (Beacon Hill I, Gardner, 1967; Thomas, 1994) dating to the Late Bronze Age (Quinnell, this volume). A further three were tentatively identified in the survey, but in an area heavily disturbed, two curved terraces may represent building platforms. It is very possible that clusters of huts have been robbed in later periods especially in areas where medieval and later activity is recorded such as at Widow's Tenement, Halfway Wall and the village area, leaving little or no evidence above ground to survey. Other individual hut circles are found amongst the relict field systems.

From Widow's Tenement to south of the village, field systems are extensive although it becomes much harder to determine a date for individual features. Re-use in the medieval period is much more likely and in significant areas, for example Middle Park and south of Quarter Wall, later ploughing in the post medieval period has obliterated or softened features. Interestingly in Middle Park walls survive east

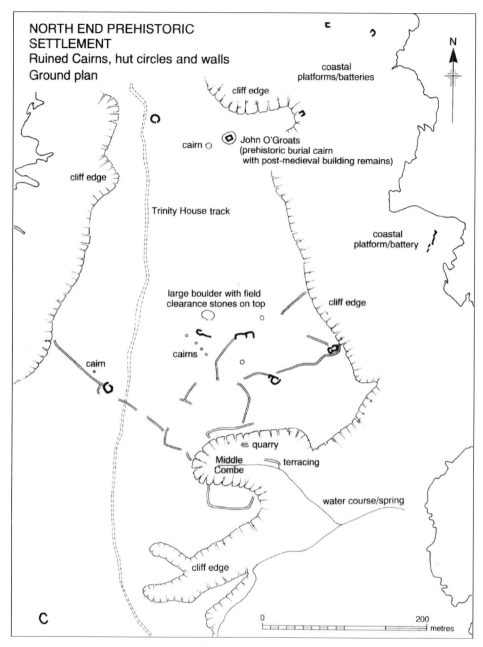

Figure 3: A plan of the North End prehistoric settlement
(National Trust Archaeological Survey; drawn by Jane Goddard)

of the main track whereas only lynchets are found to the west, suggesting perhaps that elsewhere lynchets could represent the robbed out remains of more substantial boundaries. In Widow's Tenement the discovery of Bronze Age pottery indicates settlement at this time and there may be remnants of Bronze Age structures on a platform in the angle of the north tenement wall and in a possible hut circle or other structure just to the north of the centre of the southern boundary. However the recognised field system and enclosure appear to be associated with the medieval settlement. In Middle Park, both north and south of Halfway Wall, Gardner excavated trenches into two Bronze Age hut circles. These two sites appear to be associated with the same field system. North of the wall a number of curvilinear lynchets are clear, with other lynchets extending to south of the wall, some of which are crossed by later ridged cultivation. The story here may be quite complex with reuse and development of the field system in the medieval period, perhaps culminating with the construction of the enclosed tenement identified here in the survey. In Ackland's Moor and south of the Old Light numerous lynchets and banks have been recorded. Some of these will undoubtedly belong to the Bronze Age landscape, associated with the hut circle complex at the Old Light and cemetery and sites excavated by Gardner further north. A number of field boundaries are shown on earlier nineteenth century maps (see below) and it seems likely that a prehistoric system was reused and developed throughout the medieval period and maintained until the present system was laid out in the second half of the nineteenth century. The modern enclosures south of Quarter Wall have been regularly ploughed until relatively recent times and contain few earthwork features, but as now, it is likely to have been the most favourable area for agricultural activity in the past and is likely to have contained Bronze Age fields and settlement.

MEDIEVAL SETTLEMENT AND THE DEVELOPMENT OF THE MEDIEVAL LANDSCAPE
Post Roman Period
The remarkable group of early Christian memorial stones and the number and character of identified burials on Beacon Hill, suggests a significant community or at the very least a strong Christian influence and importance of the island in the post Roman period. Perhaps this is based on a religious leader or saint as suggested by Thomas (1994). As yet we have very little evidence to suggest where or how the inhabitants were living at this time and the survey has given us no further clues. Only a few isolated, unstratified sherds of imported pottery from this period have been identified (Gardner, 1963, 23; Ray McBride in Allan and Blaylock, 2005, 88). Perhaps there is undiscovered evidence of habitation somewhere in the environs of the village yet to be located or elsewhere on the island. Perhaps the community was largely aceramic with only occasional Mediterranean imports. The discovery of the re-use of a Bronze Age hut circle at Beacon Hill (Quinnell, this volume) in the Roman period obviously indicates a long continuation of sites traditionally regarded as prehistoric.

There are a number of sites notionally designated as prehistoric but with no real evidence for their date. These include the 'Black House' excavated on the West coast of Middle Park (Gardner, 1969, 44-48); a small sub- rectangular structure and wall perched just below the plateau north of Old Light (Scheduled Monument 27633); and a partial enclosure and structure above the West Coast Fog Battery. Any of these could be of this period, but they are also in suitable locations for birding or egg collecting activities at any time in the prehistoric or medieval periods and further dating evidence is required to help with their interpretation.

Later medieval period

In the later medieval period we can begin to appreciate the benefit of written documents to expand our knowledge of life on the island and to try to make links with some of the archaeological remains. There is considerable evidence for a settled community on the island, at least for periods of time, during the later medieval period of the twelfth to fifteenth centuries both in terms of archaeological remains and historic information and events. It is likely that a farming population persisted. We have no means of knowing whether this was an indigenous, static or fluctuating population, but the island would need to be fairly self-supporting if it was not to rely on imports from across the sea. Occasional documents from the thirteenth century onward provide snapshots of ordinary life. From around 1200 rabbit warrening was established, and in 1274 it was estimated at providing 2000 rabbits a year, although we have no recognised archaeological evidence for this in the form of pillow mounds (artificial warrens) or traps. A document of 1321 mentions eight tenants paying 15s yearly, who hold land with one tenant allowed to keep the gannets; the same document tells us that there are 200 acres of waste land, used as common by all the tenants. It also states that the castle, barton and rabbit warren were of no value that year as they had been destroyed by the Scots, revealing that life was not always peaceful or easy. Collectively in these documents we are told that there was cultivated land for barley and oats, meadow and pasture, cattle, sheep and horses (Steinman Steinman, 1836, 4; Thackray, 1989, 163). It is tempting to look at the survey and see what can be identified as a tenement although it is not possible to say which sites are contemporary with each other or for how long they were in use.

Widow's Tenement

Widow's Tenement (Figure 4) is the best known of these medieval farmsteads, lying north of Threequarter Wall within its own enclosing compound. This compound appears to have been sited on a Bronze Age settlement and therefore some of the fields may have been cultivated from this time although the enclosing wall appears to be a planned unit. Within this area lies the house, excavated by Gardner (1965, 30) and approximately ten field areas delineated by walls, banks, lynchets and traces of ridged cultivation. One might assume that during the summer months crops were grown on a rotational basis in the interior and stock put out on the surrounding common land, perhaps using some of the small external enclosures as stockades when required. Wintering stock within the enclosure could allow fertilisation of the fields. The water resource appears to be an important feature, there is a dewpond, a spring in the north west corner and spring on the west side at the edge of the plateau, which is protected by side walls to produce a funnel shape in plan. This may have been a drinking water supply that needed protecting from stock.

Although seemingly isolated, Widow's Tenement has at least two neighbours (although not necessarily contemporary). The foundations of a small, roughly rectangular, structure presumed to be a medieval house and part of an enclosure wall lie to the north-east (Figure 1). This seems to have no other associated field system so, unless it was an additional dwelling related to Widow's Tenement, the inhabitants

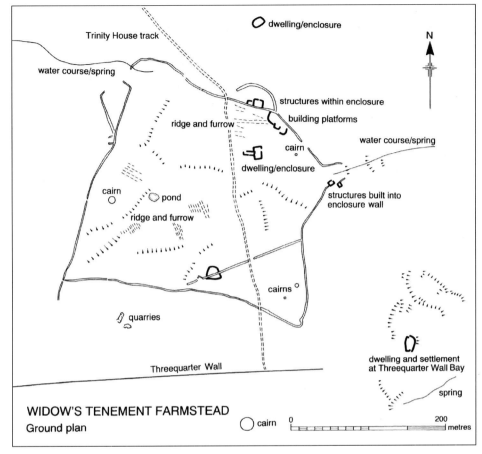

Figure 4: A plan of Widow's Tenement
(National Trust Archaeological Survey; drawn by Jane Goddard)

may have had another main occupation for example shepherding or catching sea birds. On the steep slopes to the east there lies another rectangular structure with a number of associated terraces, which may represent an independent tenement. No enclosing boundary has been identified, although individual plots of land could have been protected by banks or fences. There is a further spring to the south of this settlement and all three of these medieval settlement sites are close to Brazen Ward, which provides relatively easy access to the sea.

Settlements south of Halfway Wall

South of Halfway Wall is another well-defined tenement enclosure (Figure 5). Parts of the enclosure overlie earlier lynchets and the existence of the hut circle excavated by Gardner (1965) indicate that the tenement again overlies a Bronze Age site. This tenement has a number of similarities to Widow's Tenement; it appears to be fully enclosed, a water course is also enclosed with a small stream running from west of the main track down the east slope. Here there is also a walled funnel shape extending down the slope either side of the stream and also enclosing a small number of terraces.

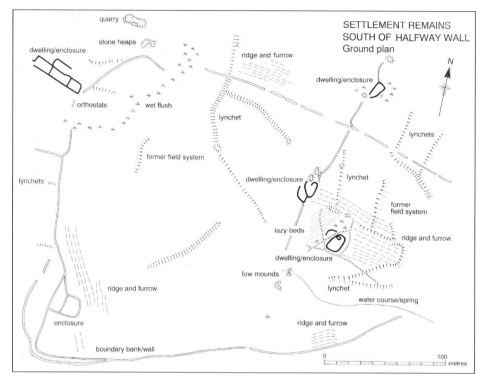

Figure 5: A plan of the medieval tenement at Halfway Wall
(National Trust Archaeological Survey; drawn by Jane Goddard)

The location of the house is problematic. During the survey it was thought that the Bronze Age house site could have been reused as it is large and sub-rectangular in shape, with what appears to be an attached cultivated plot. However, the lack of any medieval finds from Gardner's excavation seems to count against this. Other candidates may be small stone walled enclosures adjacent to the track although these could not be securely identified as buildings. Ridged cultivation is also apparent both inside and outside the walled enclosure, perhaps indicating that the tenement boundary wall was a secondary element of the medieval settlement. However, there may also have been a further settlement adjoining to the north, using the network of fields found here. There is a possible rectangular structure lying close to the track.

The quarry hospital lies within an enclosure, which is of earlier date and could be medieval. Within the wall on the north side lie the remains of a small rectangular structure and traces of cultivation ridges have been identified within the plot. There are a number of lynchets or banks to the north of this site, which may be associated with it.

Settlements south of Quarter Wall

Earlier nineteenth century maps may give us an insight into the appearance of the medieval landscape south of Quarter Wall. These are usefully described by Langham (1991) in his attempt to identify the location of 'New Town'. These maps were drawn up before the existing system of enclosures was developed. The organic looking field systems represented could have medieval or earlier origins and it is instructive to compare these with the relict field systems recorded in the survey. The most useful map is that compiled from surveys undertaken by the Ordnance Survey to produce the 1 inch to 1 mile map, and drawn up at a scale of 1:10,650 or 6 inches to 1 mile. This is a compilation of three surveys; initially by Thomas Compton in 1804, corrected by A.W. Robe in 1820 (Figure 6) and with additions and corrections made by Lieutenant Denham in 1832. A map included with auction papers from 1840 is similar (from a copy of a tracing made by Tony Langham). The easiest enclosure to pick out from the survey is an irregular oval or sub-rectangular enclosure in the central area now cut by

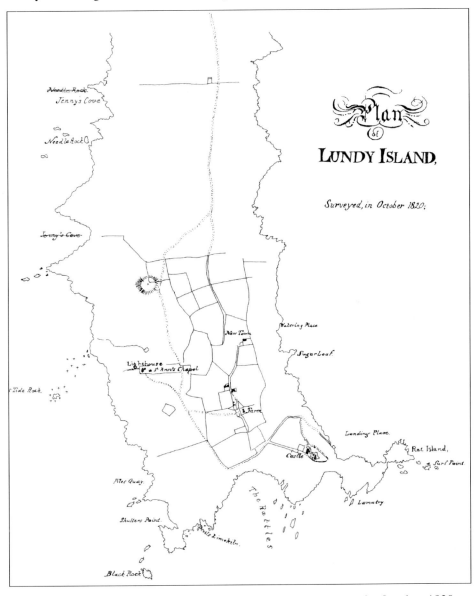

Figure 6: Map of Lundy surveyed by the Ordnance Survey in October 1820
(A.W. Robe)

the Old Light Wall. This also appears to be an early enclosure within the field system perhaps even prehistoric in origin. A number of other boundaries run up to or lead away from it. To the north there appears to be a broad track or drove way, which approximates to the line of the shape of field boundaries shown on the nineteenth century maps although no track is shown. The track can be followed on the ground as an earthwork feature leading into Bull's Paradise. Running along the southern boundary of this enclosure another possible track also leads out of 'Bull's Paradise', this is shown on the map of 1840. A large lynchet roughly parallel to this southern boundary also appears to be shown in part on both maps mentioned above. To the north, this relict field system is crossed by the present line of Quarter Wall. There are traces of ridged cultivation both just to the south of Quarter Wall and south of the air strip. An outlying field known as 'Friar's Garden' is depicted on both maps and was also recorded by the survey. It may be that this enclosure survived for longer that its associated field system or was reused as there are further lynchets and banks recorded in the survey to the north-east but not shown on any maps. No traces of 'New Town' and the fields and trackways shown in this area on the maps, were found by the survey (now Tillage Field) indicating fairly intensive clearance and ploughing from the later half of the nineteenth century onwards, although it is possible that negative features such as ditches and robbed out wall foundation trenches will survive here.

From the nineteenth century quarries southwards, short sections of terraced walls have been constructed on the steep slopes of the east sidelands, occasionally with evidence for a small associated structure. The logical interpretation of these is as garden or cultivation plots. Dating them is not easy. Some may be associated with the construction of the nineteenth century gardens and paths leading up the east side from Millcombe. In origin however, they are probably earlier. This is the warm and sheltered side of the island and it may have been utilised in earlier periods for a variety of types of crop cultivation, possibly including managed trees. Access to the plots from the village across the fields or from 'New Town' may have been easier than it is today.

The focus of the medieval settlement lies around the site of the present farm buildings and the fields to the west and south known as Bull's Paradise and Pigs Paradise. Earthworks have long been recognised here and chance archaeological discoveries, encouraged a number of small-scale excavations, most recently in the 1960s. From the mid-twelfth to the mid-thirteenth century the island was held by the Marisco family. Gardner suggests that the foundations of a substantially built structure in 'Bull's Paradise' partially excavated in the 1960s, just to the west of the present farm buildings, was their stronghold (Gardner, 1963, 1969). Here, a courtyard containing a 'waterhole' and lean-to structures, was enclosed by a heavily built wall and surrounding ditch. Burials, with a building in close proximity and the discovery of part of a piscina suggest that there was also a chapel and cemetery here which had gone out of use by the seventeenth century (Gardner, 1963). A watching brief of service trenches in Pigs Paradise in 2000 (Allan and Blaylock, 2005) recorded what appears to be part of a medieval farmstead, including a possible building, wall and a small number of pits and post-holes. Further structural walls were found under 'Quarters' when these buildings were constructed in 1972. The quantity of pottery excavated in 2000 allowed a detailed examination to be undertaken. This included petrological study of the inclusions in the pottery and chemical analysis of the pottery fabrics. Of the 1451 sherds retrieved, the vast majority (80%) were from North Devon Coarseware vessels, largely unglazed cooking pots probably imported from Barnstaple and Bideford, with a smaller number of coarsewares probably coming from Exmoor. Some glazed fragments, principally jugs, and some sherds of unglazed cooking pots, are from Ham Green, near the Bristol Avon. Six sherds of glazed ware from Redcliffe, Bristol, were also present with a small collection of limestone-tempered wares from Wiltshire including a tripod pitcher, two cooking pots and a jug. The pottery is suggestive of a significant farmstead or hamlet settlement at sometime between the mid twelfth to mid fourteenth centuries, with prime market links in Bideford or Barnstaple, but also links to Exmoor and Bristol and possibly South Wales. The mention of eight tenants in 1321, the other known sites, and the quantity of archaeological evidence from this area suggests that there was more than one tenement, more likely three or four, centred on the village at this time. It is likely that at least one of these lies in Pigs Paradise.

Castle

After the capture and execution of William de Marisco for plotting against King Henry III, the island was brought directly under royal control, and the first castle was built in 1244 in an imposing position above the Landing Bay (Ternstrom 1994). The surrounding rampart and ditch comprise the most impressive earthwork remains from this period, although the present structure is likely to have been heavily rebuilt from the seventeenth century onwards (Figure 7). The castle was essentially a strong keep within these defences, with a gatehouse on the landward side (Figure 8).

CONCLUSION

In conclusion therefore it seems likely that in the later medieval period, roughly the twelfth to fifteenth centuries, there was a village or hamlet in a similar position to or slightly north of the present village, comprised of perhaps three or four tenements or farmsteads and at least in the later part of this period, a cemetery and probably a small chapel. The hamlet is likely to have been surrounded by enclosed fields with a number of tracks to provide access to the fields and to rough pasture or common land beyond. If one accepts Gardner's thesis, the hamlet would have been dominated by the defensive compound discovered in 'Bull's Paradise', until the construction of the castle in the thirteenth century took the administrative focus out of the village to a more strategic position above the Landing Bay. Away from the village there were a small number of outlying farmsteads or tenements; with at least Widow's Tenement, and the tenement at Halfway Wall, lying within their own enclosed fields protected from stock by an enclosure wall, outside of which was common land. Other settlement sites appear to represent smaller units, the largest of which is the medieval house on the line of Threequarter Wall situated among terraces, which could have been protected from stock by banks or hurdles. Those smaller still at the site of the quarry hospital and north of Widow's Tenement could have derived more

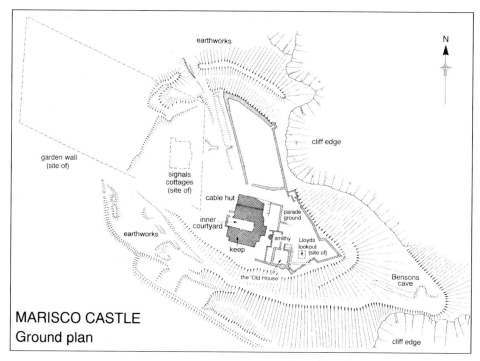

Figure 7: A plan of the Castle, Lundy
(National Trust Archaeological Survey; drawn by Jane Goddard

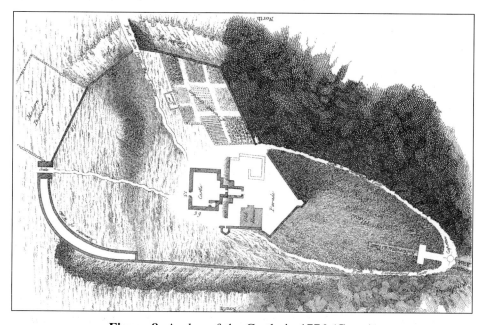

Figure 8: A plan of the Castle in 1776 (Grose)

of their income from activities such as fishing, taking sea birds and their eggs or possibly rabbit warrening.

As mentioned in the introductory section, this paper only presents our thinking on certain elements of the survey. I have not even touched on a number of important aspects of the archaeology on Lundy such as coastal lookouts or defences, neither have I attempted to discuss the considerable archaeological remains relating to the post-medieval period, many of which are readily appreciated in the landscape by Lundy's many visitors. These were also recorded in detail by the survey and the work of Myrtle Ternstrom in studying the historic documents from this period, has added specific detail on subjects and issues that we can only guess at for the earlier periods. Never the less I hope this paper has provided an overview of the fascinating early landscape of Lundy of which there is still much to be understood. One of the biggest lessons learnt in studying Lundy's past is that although it may seem so when coming in to land by helicopter, Lundy's landscape is far from flat.

ACKNOWLEDGEMENTS

The biggest debt of gratitude is owed to Caroline Thackray who for many years led National Trust surveying expeditions to Lundy. Her extensive research has been drawn on and ideas for overall interpretation developed from her lively discussions with myself and other team members. I am very grateful to her for discussing the content of this paper. Thanks are also due to Henrietta Quinnell and Myrtle Ternstrom for commenting on a draft of this text. Very many people have contributed to helping us develop our understanding of the island, most recently John Allan, Henrietta Quinnell and Ann and Martin Plummer in their studies of material remains. Valuable discussion and useful information have also been provided by Myrtle Ternstrom, Keith Gardner, Vanessa Straker and Stuart Blaylock together with other friends and colleagues. The jigsaw pieces of our knowledge and interpretation have been brought together by many, including the staff and volunteers of the landscape survey, all of whom gave so much in the recording and interpretation of the island, a most particular mention of the contribution of Gary Marshall, Jeremy Milln and Angus Wainwright, should be made.

REFERENCES

Allan, John, and Blaylock Shirley, 2005. Medieval Pottery and Other Finds from Pigs Paradise, Lundy. *Proceeding of the Devon Archaeological Society*, 63, 65-91.

Gardner, Keith S., 1963. Archaeological Investigations on Lundy 1962. *Annual Report of the Lundy Field Society 1961*, 15, 22 - 33.

1965. Archaeological Investigations, Lundy, 1964. *Annual Report of the Lundy Field Society 1963-64*, 16, 30-32.

1967. Archaeological Investigations, Lundy 1966/67. *Annual Report of the Lundy Field Society 1965-66*, 17, 30-33.

1969. Lundy Archaeological Investigations 1967. *Annual Report of the Lundy Field Society 1968*, 19, 41-48.

1972. *The Archaeology of Lundy:* A Field Guide. Bristol: Landmark Trust.

Langham, A.F., 1991, Newtown, Lundy: A lost Georgian Settlement. *Annual Report of the Lundy Field Society 1990*, 41, 55-64.

Steinman Steinman, G. 1836, *Some Account of the Island of Lundy*. From the Collectanea Topographica et Genaelogica Vol IV, reprinted 1947.

Ternstrom, Myrtle, 1994. *The Castle on the Island of Lundy, 750 Years 1244-1994*. Stroud: Alan Sutton.

Thackray, Caroline, 1989. *The National Trust Archaeological Survey, Lundy*. Volumes 1 and 2, unpublished.

1997, The Archaeology of Lundy, in R.A. Irving, A.J. Schofield and C.J. Webster (eds), *Island Studies Fifty years of the Lundy Field Society*, 67- 76. Bideford: The Lundy Field Society.

The National Trust, 2000. *The Archaeology of Lundy*, leaflet.

2002. *The Archaeology and Landscape of Lundy*. A Field Guide.

Thomas, Charles, 1994. *And Shall These Mute Stones Speak*. Post Roman Inscriptions in Western Britain, 163-182, University of Wales Press.

LUNDY'S HISTORY: THE COURSE OF CHANGE

by

MYRTLE TERNSTROM

6 Queensholme, Cheltenham, GL52 2QE
e-mail: mst@waitrose.com

ABSTRACT

The history of Lundy is examined by the changes through the historical time frame, with reference to the maps that register them. Two broad agents of change are cited: internally the effects of the attitudes of the island's successive owners, and the influence of external economic and social changes affecting the country as a whole.

Keywords: *Lundy, history, maps, island owners*

INTRODUCTION

The first depiction of Lundy, other than by rather vague outlines, was not until John Donn published his map in 1765 (Figure 1). The latest map we have is that which the Landmark Trust provides for visitors, (Figure 2) and this talk will try to trace the path from one to the other.

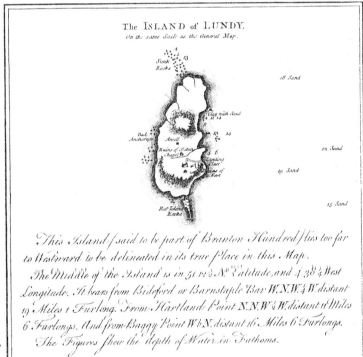

Figure 1: Map by Benjamin Donn, 1765. National Archive, WO 78/5675

BEFORE DONN

The fort, or castle, was built in 1244 after Henry III had captured and executed an outlaw sheltering there, William de Marisco (Figure 3). The king then kept Lundy in his own hands. This tells us that Lundy was a bastion, indeed a natural fortress that could be used either by rebels and enemies, or for the defence of the Bristol Channel. The Marisco family held Lundy - on and off - from the eleven hundreds until 1344 (Ternstrom, in preparation). It is interesting that the castle was funded in part by the sale of rabbit skins. (CCR 1243) The Isles of Scilly and Lundy provide two of the first records of rabbits, islands being particularly well suited for warrens as they prevented escape. Both the skins and the meat were very valuable assets at that time (Veale, 1957).

Figure 3: Execution of William de Marisco, 1242, from *The Drawings of Matthew Paris* edited by M.R. James. By kind permission of the Walpole Society

Excavations of medieval settlements (Figure 4; Thackray 1989), together with inventories dated 1274 and 1321, (Steinman, 1836) a deed of lease made between 1182 and 1219 (DRO), together with Holinshed's description (*c*.1586), all indicate that the island was well populated and prosperous in medieval times. This probably resulted from sheep rearing when the wool trade flourished in Devon. Another resource for the islanders, of importance into the mid-nineteenth century, was the seabirds that crowded the island in the breeding season so that the birds,

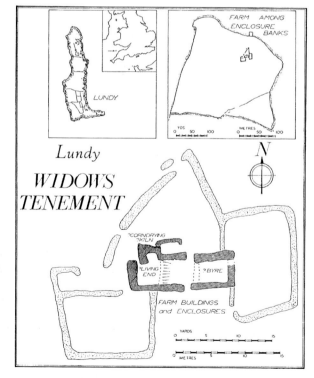

Figure 4: Plan of Widow's Tenement. By kind permission of K.S. Gardner

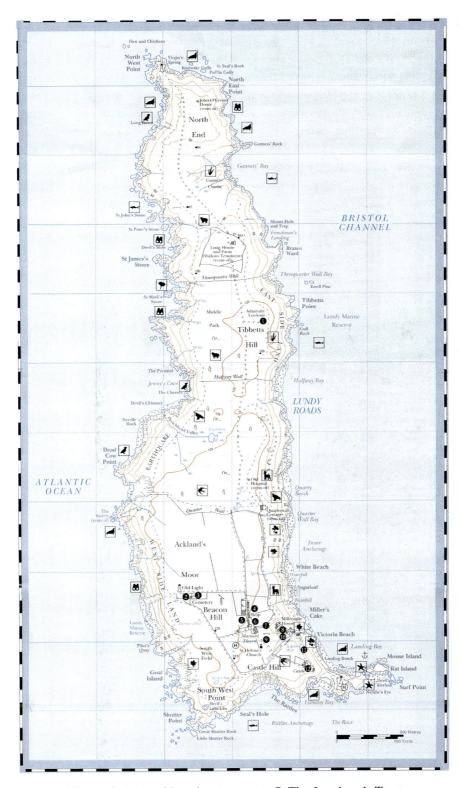

Figure 2: Map of Lundy at present. © The Landmark Trust

their eggs and feathers were all harvested and sold. A limited trade in Lundy peregrines, which have been highly esteemed, continued until the first world war.

Bevill Grenville, grandson of the famous Richard Grenville of the *Revenge* who had acquired the island in about 1577, was given the island on his marriage in 1619 and he fell in love with it. He had great plans to improve it for farming, fishing and breeding horses, but was overtaken by debt, and lost his life in the Civil war in 1643 (Stucley, 1983). It is thought that he and his father constructed defensive platforms around the island coasts, which were necessitated by piracy and by the Spanish threat to protestant England. The most important of these is Brazen Ward, (Figure 5; Gardner, 1971) which was well fortified, as it allows of landing. It is also proposed that he built two houses in front of the castle, as by then it was ruinous (Figure 6; Ternstrom, 2000).

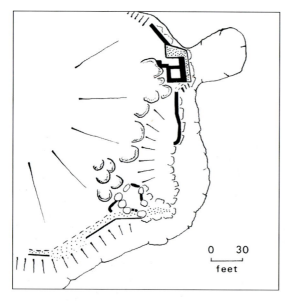

0   30
feet

Left: **Figure 5**: Plan of Brazen Ward. By kind permission of K.S. Gardner

Below: **Figure 6**: House on the Castle Parade, from N.W., excavated 1985. *Photo: M. Ternstrom*

Except for a period during the Civil War, the Grenville family and their descendants owned the island until 1775. (Ternstrom, 1998) During this time agents collected the rents from tenants, who had no money or inclination to invest in it. Thus Lundy itself went through a period of considerable neglect until it was sold in 1775, although there is evidence that great profits were made by smugglers. In 1750 Lundy was leased to Thomas Benson, a shipping merchant and MP for Barnstaple, who concealed smuggled tobacco there. He had also contracted to transport convicts to America, but instead landed some of them on the island where they were used as a slave work force (Thomas, 1959).

Benson's career on Lundy came to an end with the exposure of his having used it to carry out a shipping insurance fraud. He fled to Portugal, leaving the captain of the ship involved to hang for the crime. Although the cave below the castle is named after him, markings scratched on the interior walls show that it was extant before his occupation of the island (Figure 7). Benson is one of the more colourful characters associated with Lundy, but his interests lay in smuggling, which continued to be rife. In 1783 one ship tried to escape from Lundy that, when captured, was found to have on board 7,000 pounds of tea, 2,200 gallons of brandy, and 823 gallons of gin (National Archive, 1783).

Above and left: **Figure 7**: Benson's Cave, below the castle. Entrance and interior. *Photos: R. Derek Sach*

DONN, PARKYAS AND THE 1819 SURVEY

Donn's map reflects that even as late as the mid-eighteenth century very little was known about Lundy. It was remote, and lacked anything of interest that would induce the traveller to face the hazards of getting to the island and back again. The marking of rocks, the depths of waters, and 'Bad Anchorage' indicate that the map would have been of most use to seamen. The chapel on the Celtic burial enclosure is named St Ann, as it is in all the early maps. It shows the 'remains of a fort,' with flag flying, and a path leading - presumably - to it.

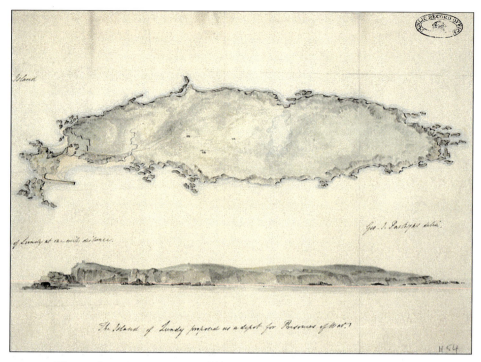

Figure 8: Map by Geo. Parkyas, 1804. National Archive, MPH/54

A map drawn by George Parkyas in 1804 (Figure 8) at first appeared to be a mistake, as it shows a pier at the south end of the landing bay, and we knew that one had been started but never completed. But research leading to a letter to the War Office, which accompanied it, shows that he had visited the island in 1775, when the pier was under construction, and had assumed that it had been completed in the mean time. The purpose of the map and letter was for the then new owner to sell the island to the government for the detention of French prisoners of war, in expectation of a huge profit. But the sale did not materialise (National Archive. 1804/WO).

The pier had been started as part of island improvements made by John Borlase Warren, who bought the island for a gentleman's country estate in 1775. The extent of his works can be seen in a drawing made in 1819 for Trinity House (Figure 9). This shows that the path from the landing place led up to the castle, where there was a farmyard, and from there a path led to the top of St John's Valley. There is a farmhouse, and the Quarter Wall had been completed, with field enclosures and 'New Town' to the south of it. Warren's plans also included a fine new residence and other buildings, which were never achieved as he ran into such debt that his trustees sold the island in 1803 (Figure 10; Ternstrom, 1999). The drawing for the pier is of additional interest as it shows the remains of the 'old pier' at the same site, which is the only indication that Bevill Grenville had carried out his intention in 1630 (Stucley, 1983).

Right:
Figure 9:
Sketch map
1819. By kind
permission of
Trinity House

Below:
Figure 10:
Design for a
residence for
Sir John
Borlase
Warren.
M. Ternstrom
collection

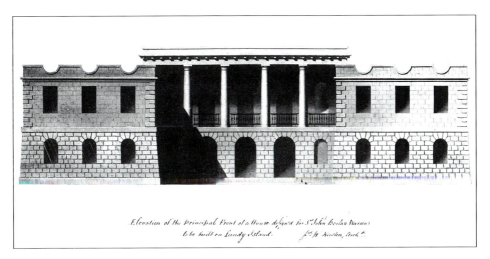

TRINITY HOUSE

The rate of wrecks, and losses of vessels and men, resulted in Trinity House taking over the building and care of lighthouses, whereas many had until then been in private hands. Although Lundy was a sea marker by day, fogs were not infrequent, and as there was no illumination at night, the rocky coast was a hazard to shipping. In 1820 the lighthouse was completed on Beacon Hill - the highest point of the island (Figure 11; THGM, 1819-20). Trinity House was the first external institution to gain rights on the island, and had the first of a series of leases and concessions in the nineteenth century that led to the map of Lundy as it is today. Apart from the church, it is also the only one of these institutions to retain their rights in island buildings and rights of way.

Figure 11: Design of the lighthouse, 1820. Daniel Alexander. By kind permission of Trinity House

The Trinity House map of 1820 (Figure 12) shows the lighthouse and its compound, and also shows the castle as an inhabited building, and that three fields to the west of it were enclosed and cultivated. (THEA, 1820) If Warren had intended the enclosed fields south of Quarter Wall for arable land, it was probably then that the Halfway Wall was built to enclose more land for pasture.

The coming of the lighthouse opened Lundy to greater contact with the outside world, with deliveries from Trinity House ships and regular visits from its officials. It also added two families, who were not employed by the owner, to the small population. It might have been thought that they would be welcomed, but for the first few years the two were at loggerheads (Ashley, 1841). It is also seen on this map that Trinity House built a quay, and a cart road up the side of St John's Valley to the lighthouse, which was needed for the carriage of heavy supplies of oil and coal. All goods still had to be brought from the landing place up the old steep path to St John's Valley either by sleds, ponies, or manpower (Figure 13).

Figure 12: The south end of the island in 1820.
By kind permission of Trinity House

Figure 13: Painting by Dominic Serres for Borlase Warren, 1775, showing the
original steep path. Photoprint by kind permission of K.S. Gardner

Trinity House was reluctant to concede that the magnificent tall lighthouse suffered the enormous disadvantage of being obscured from time to time by high-level fog. The lantern was enlarged and the beam intensified in 1842, and again in 1857, but without overcoming the basic problem (Ternstrom, in press). So in 1862 they attempted to remedy matters by building a fog signal station low down on the cliffs at the western side of the island (Figure 14; THGM, 1861-2). It was furnished with two cannons, and two cottages were provided for the families of the gunners, making four Trinity House families on the island in all. The gun house, the cannons, and the remains of the cottages are still there, now called the Battery, and they make a wonderful sunny and sheltered spot to enjoy the scenery in fine weather.

Figure 14: The Fog Signal Station from the north. Heaven collection

Trinity House was eventually forced to acknowledge that the lighthouse on Lundy, even supplemented by the fog signal station, was not satisfactory. In 1897 these were replaced by new lighthouses at low level at the north and south ends of the island, and the Old Light and Fog Signal Station reverted to the owner of the island. The new lighthouses were classed as rock stations, so that the Trinity House families left the island (THGM, 1895-7).

ORDNANCE SURVEY

The first Ordnance Survey map was also published in 1820. It gives more names, notably Tibbets Hill, and Johnny Groats House at the North End. This was built over an ancient burial cairn at the high point, which suggests that it was a watch house, possibly built during the wars with the French, and mentioned as such by a writer in 1776. (Grose, 1776) Of the named coastal features all but three are still in use today, which reflects their consistent use as sea-markers. The map also shows, for the first time, the track to the North End. Originally 'North End' or 'North Part' referred to the land beyond Quarter Wall (Figure 15). It now refers to the part of the island beyond the Threequarter Wall, where it has at least twice suffered burning to the bare rock, the last time in the 1930s (Gade, 1978).

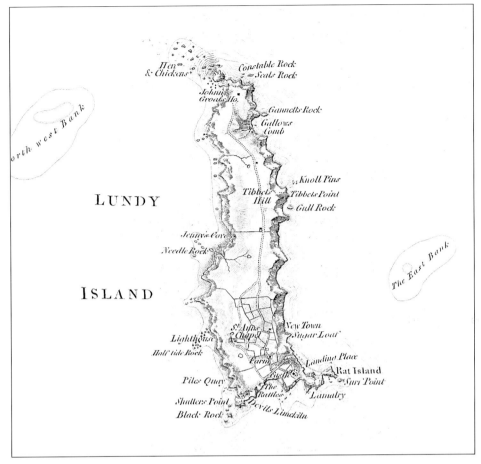

Figure 15: Reproduced from the 1820 Ordnance Survey map.
Crown Copyright

SALE MAPS

The island was for sale from 1818, and the advertisement for the auction is the first recorded occasion when Lundy was stated to be exempt from tithes and taxes (BL, 1822). In 1822 Sir Vere Hunt found a buyer, who appeared to be on the risky side of eccentric, but insisted on proof of the exemptions and refused to accept an indemnity (Limerick, T22). Such proof could not be found, because these privileges rested on custom rather than any legal ruling, and arose from the remoteness of the island and the lack of anything of value that would have made it worth a tax-gatherer's efforts. The case went to court, and the vendor lost. The appeal of the exemptions would have rested on the fact that income tax of two shillings in the pound had been imposed, and the land tax was four shillings per acre. The matter of the tithes was not in question as there was then no church on the island.

For the sale in 1822 J. Wylde prepared a map in which many field names and coastal sites are given (Figure 16; BL, MSS) Overall the pattern of field systems is the same, but of the field names not one is in use today, which indicates the lack of

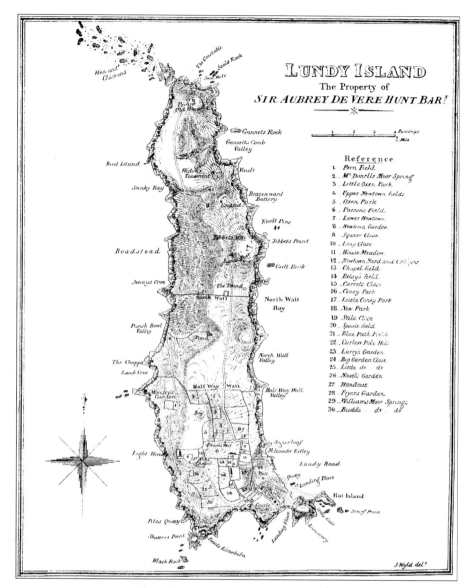

Figure 16: Map for the sale of Lundy in 1822 with details of field names.
By permission of the British Library, ADD 40345 a: MAPS 299A

continuity in Lundy's population. This is not simply a matter of names: the diminution of knowledge, experience, and commitment are concomitant.

THE ANTIQUARIANS

Two early significant contributions were made to the Lundy bibliography: the first by Francis Grose in *The Antiquities of England and Wales,* 1776, which gives an outline of the history so far as it could then be traced from reference to the Rolls, with some evidence from a long-term island resident. An important and much appreciated set of engravings was included, with two views of the castle, and a plan (Figures 17-19).

Figure 17: View of the castle from the N.E., 1775. F. Grose, 1776

Figure 18: The castle from the N.W. 1775. F. Grose, 1776

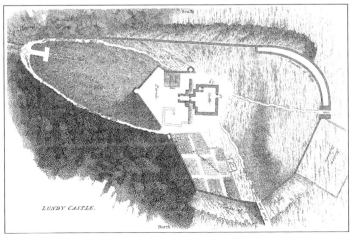

Figure 19: Plan of the castle, 1775. F. Grose, 1776

The castle had been rebuilt by Thomas Bushell, another intriguing Lundy character, who was governor of the island for the king during the Civil War (Bushell, 1647). It is interesting that Grose makes the first reference to the church as St Helena's, 'very small and ruined to the foundations.' It is assumed that this refers to the chapel in the burial ground at Beacon Hill, although a late medieval burial ground in the area of the present Bulls' Paradise has been described by Keith Gardner, including the foundations of a building thought possibly to have been a chapel (Gardner, 1962).

The second account of Lundy was given by G. Steinman Steinman FSA in 1836, which greatly extended the references to the Rolls and gave an accurate history to that date. It remained the basis from which subsequent historians have worked, and the text is so interesting that Mr Harman reprinted it, privately, in 1947.

THE HEAVEN ERA

In 1836 William Hudson Heaven, of Bristol, bought Lundy with the intention that it would be a summer resort for his family, where he would be able to enjoy the shooting (Heaven family papers). No doubt the allure of an island fiefdom, together with the traditional exemptions, were part of its appeal. At the time of Heaven's purchase Lundy was in essence a farm with a lighthouse and a castle. There was no church, no school, no doctor, no shop, no meeting room, and it was extra-parochial. Whoever took employment there depended entirely on the owner.

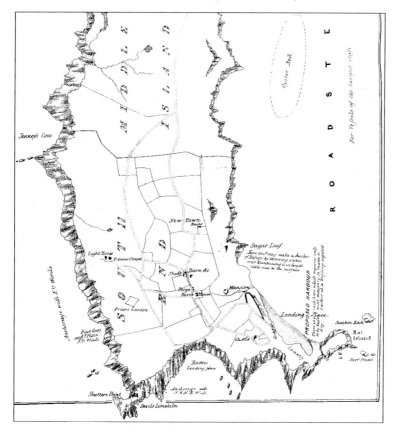

Figure 20: The south end in 1840. Heaven collection

- 44 -

VIEW OF RESIDENCE ON THE ISLAND.

Figure 21: The Villa, 1840. Heaven collection

Heaven set about making the island a suitable place for a gentleman's family, and his map of 1840 (Figure 20; NDRO) points to the works that were carried out. A delightful villa was built at the head of Millcombe Valley (Figure 21) and a road - or rather a track wide enough for carts - was constructed from the quay, through Millcombe, to bend back and meet the Trinity House track at the present Battlements. This meant that heavy loads could be carried from the landing place in carts, and the ladies of the family could ride in the carriage. The farmhouse was extended and rebuilt, and the interior of the castle adapted to make dwellings for labourers, as his predecessors had also done.

Unfortunately the map was made in an attempt to sell the island because of the virtual collapse of Heaven's previously ample income. The map is based on the 1820 OS, but with an estate agent's fanciful embellishments. In the superscript Brazen Ward is marked as 'The rock from whence granite may be exported by merely laying down moorings.' At the landing bay '... the govt may make a harbour of Refuge by throwing stones over it and continuing it in length till it rises to the surface,' and 'Proposed Harbour: There is a rock here which requires only to be toped (sic) with masonry to make a dry harbour at trifling expense.'

Heaven had paid a high price for the island (almost £10,000) and could find no buyer from whom he might recover his outlay, and despite further attempts to sell, the family kept the island until the end of 1917. As Grenville and Warren had found before him, investment in Lundy did not bring a consequent improvement in income, but it was necessary to support it with independent funding. What was exceptional in the history of island ownership was that from 1851 Lundy was the Heaven family's home. For the first time there was a resident squire. There were extensive walled gardens for produce, plantations of trees, flowers and shrubs, and the commercial exploitation of the seabirds was forbidden.

THE GRANITE QUARRIES

Heaven had cast around for sources of income in the possible exploitation of the granite, and in a search for minerals. Copper and other minerals were found, but not in such quantities that would repay the costs of production and transport - and, in any case, he did not have the necessary capital. But the rapid surge in the construction of public works meant there was a demand for building stone, and this, combined with the Companies Act of 1862 that limited the liability of investors to their own share-holding, combined to cause others to take a more optimistic view of the possibilities of Lundy granite.

A lease was granted in 1863 to a Mr McKenna for the Lundy Granite Co. to begin operations. A site was chosen on the sheltered East Side, and great changes for Lundy followed. The population was swollen by about 200 workmen and some of their families. The company built a quay and jetty near the works, with an inclined railway system to move granite down to it, and supplies up to the plateau. For the first time Lundy had a shop - the Store - which was also a refreshment room. For this they built a north wing to the farmhouse, where there was also a bakery and a cottage for the store keeper (Figure 22). Three beautifully sited cottages were built for the managers, now the ruined Quarter Wall Cottages, with a row of cottages in the High Street that are now called Barton cottages, and three other rows of cottages north of the Quarter Wall of which only the foundations remain (Ternstrom, 2005).

For the first time there was a doctor on the island, with a small hospital, of which the ruins still stand (Figure 23). An iron hut was erected for a 'Mission Room' which met the need for a parish hall for meetings, services, a schoolroom, and

Figure 22: The Store, bakery (at right) and store-keeper's house. Heaven collection

lantern shows. Another wing was added to the south end of the farmhouse, barracks were put up opposite the Barton cottages, and a time-keeper's hut at the top of the path down to the quarries themselves. Of these three the time-keeper's hut remains, and has been repaired (Figure 24). There are four quarries, and the outlines of the work have been softened by vegetation, so that it is now a sheltered and favourite place for walks and picnics. J. R. Chanter's book, *Lundy Island*, published in 1877 following his paper in 1871, was the first monograph on Lundy and gives a map (Figure 25) that shows all the sites of the granite works. The publication of the book had a stabilising influence on island names and versions of its history.

Although the management of the Quarry Company was disastrous and resulted in its liquidation in 1868, the legacy was the shop and bakery, twenty-six cottages, and a fund of cut granite. Anything that could be removed had been sold, but all the buildings reverted to the owner, which meant that the previous sparse and Spartan accommodation for labourers could be improved upon.

The rapid rise in the volume of shipping following the development of the steam engine and large iron vessels, together with the enormous expansion of trade in the

Figure 23: Ruins of the Quarry Hospital. Privy at right, and fireplace diagonally opposite. Entrance was from west. *Photo: R. Derek Sach*

Figure 24: The time-keeper's hut in 1971 *(above)* and now *(below)*. *Photos: R. Derek Sach, M. Ternstrom*

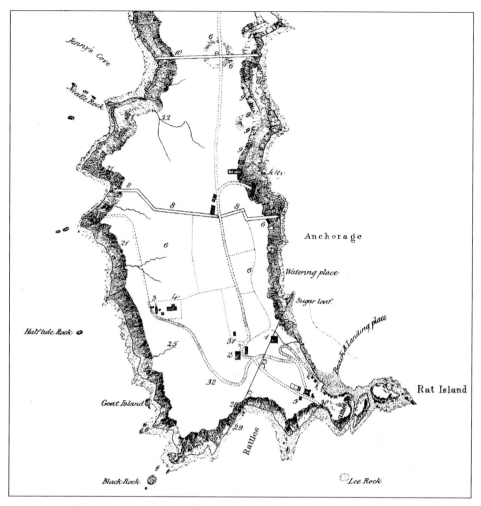

Figure 25: South part of the map from Chanter's book, 1877

nineteenth century, meant that owners and insurers were anxious to receive news of their vessels. It was asserted that the Bristol Channel was used by one-sixth of the nation's shipping, and accordingly it was desirable that there should be a means of communication. In 1884 Lloyd's negotiated the lease of a site near the castle for a flag signal station and built a pair of ugly suburban-style cottages to house their employees (Figures 26, 27; Lloyd's, 7778).

There was also a very important submarine cable for the telegraphic transmission of messages. Its use was not strictly confined to Lloyd's but meant that urgent telegrams could be sent and received without delay, instead of waiting for the next convenient ship. News was received, the correct time could be established, and medical help summoned without the need to light a beacon, to wait on a ship to take a message, or risk a crossing in a small and slow boat. One man with his family was posted to the station, and a dawn to dusk watch was kept. The average number of reports sent annually was 700, which does indicate the value of this station to shipping.

Figure 26: Lloyd's Signal Cottages in 1920. *Photo: H. Jukes, LFS archive*

Figure 27: Lloyd's Signal Hut with the flagstaff. Heaven collection

Figure 28: GPO Cable Hut 1893. Heaven collection

Between 1888 and 1893 the cable was out of action until the GPO replaced it, and built the cable hut against the north wall of the castle. This building has now been extended and converted into a small letting cottage that has wonderful views along the east coast (Figure 28).

Heaven contracted for a boatman to serve the island, who also carried the post from the post office at Instow, but when the island was leased to a Mr Wright in 1885, and the Heaven family retained only the reserve area they had fenced off during the granite company operations, he arranged for the GPO to service the mails. A sub-post office was set up at the island store, with an islander as postmaster, and thus one more facility and mainland authority was represented on the island, and remained so until 1928 (Figure 29).

LANDING THE WEEKLY MAILS ON LUNDY ISLAND.

Figure 29: The postmaster with the GPO mailbag brought by the island boat, *Lerina*, in the background. Myrtle Ternstrom collection

After Heaven's death in 1883 his eldest son, the Reverend Hudson Heaven, took over the island. He had lived on the island since 1864 to administer to the religious needs of the population, but was often of weak health and resolution. In 1885 a relative gave funds to erect a pre-fabricated chapel with a separate Sunday school (Figure 30). The Bishop of Exeter dedicated the chapel, and from then on Lundy was part of the diocese of Exeter, although it was neither a parish in itself, nor part of any other parish, and Heaven's title of vicar was by courtesy. The chapel was demolished in 1918, but the Sunday school has been refashioned into the much-favoured Blue Bungalow (Figure 31).

During the Reverend Heaven's time the Threequarter Wall was built to extend that part of the island for use for pasture. This, with the wall built by his father in 1838 across to the lighthouse, completed the present division of the island into four sections.

There are several records of the islanders having been involved with the rescue or care of the survivors of wrecks (Ternstrom, in press), but the rescue of the crew of the *Tunisie* in 1892 was an exceedingly difficult and exhausting one, carried out in very bitter weather. It resulted in the Board of Trade's acting to establish a Coastguard-trained life-saving company among the islanders, equipped with rockets to carry a line to a stranded vessel (National Archive, 1892). A hut was built to house the cart and apparatus, and a rocket practice pole was put up at the south west

Figure 30:
The small
church built in
1885. Heaven
collection

Figure 31:
The former
Sunday School,
now with
annexe and
painted blue.
*Photo: Myrtle
Ternstrom*

field (Figures 32, 33). The pole remains, sadly without its accoutrements, and the hut is now used for an island history information centre and display.

In the present age it is sometimes difficult to grasp the intensity of religious fervour among some sections of the population during the nineteenth century. When the Reverend Heaven received a legacy, and despite the debts and poverty that the family suffered, he used it to construct the large stone-built church for which he had long cherished an ambition (Figure 34). This has since been much criticised as unsuited to Lundy in both style and size, but perhaps the best that can be said of it is that it is of its time. To him it was a family memorial, a manifestation of faith, and conferred on the island what had long been missing: a consecrated building where all the rites of the church, including marriages, could be conducted. It stands proud on the horizon, a sign to seafarers who, for him, formed an important part of his mission on the island.

Figure 32: The Rocket Shed in 1922. *Photo: The late A.E. Blackwell. Myrtle Ternstrom collection*

(Above) **Figure 33**: Practice with Rocket Life Saving Apparatus at Rocket Pole. Heaven archive

(Left) **Figure 34**: Interior of the church built in 1897. Heaven archive

THE CHRISTIE OWNERSHIP

After the death of the Reverend Heaven in 1916 the heir, Walter Heaven, was penniless, so at the end of the following year Lundy was sold to A.L. Christie of Tapeley, Instow. Both the farm and the buildings were by then in a very severely run-down condition (NT, 1915-21). Christie has hitherto received short report and no credit for bringing about the recovery of the island. Whereas the Heavens had extrapolated the exemption from tithes and taxes to claim Lundy's total independence from mainland authority, Christie was a man familiar with the management of property who was not interested in such claims. He had a thorough survey made by a consulting engineer, and put the management of the island into the professional administration that could make it a productive part of his estates (NT, 1918-21; NDRO, 1918-25).

The water system was restructured - including the encasing of Golden Well and the construction of the leat across the Common in front of the church. For the first time on record the owner of the island bought a boat, and to facilitate landings a slipway was built at the Cove, where Grenville and Warren had built their piers. The engineer also constructed what are now, mistakenly, known as the Montagu steps to allow for

Figure 35: Montagu Steps, constructed in 1920 for West Side landings. Myrtle Ternstrom collection

(precarious) landings on the West Side (Figure 35). The buildings were repaired, the farm rehabilitated and re-stocked, and in 1920 the whole (except the Villa) was let to the very competent management of Mr C. H. May. Shortages of materials and labour during the war had prevented the construction of the planned harbour, and afterwards the costs had risen to such an extent that it was impossible (NDRO, 1920-25).

It seems sad that Christie was reputed have been motivated by the wish to possess all the land he could see from his Instow estate, and appears to have taken no pleasure in the island. He seldom visited it, neither was Mr May resident. However, experienced staff were put in charge of the farm and - for the first time - a hotel, for which the farmhouse was adapted. No interest was taken in the so called privileges of Lundy, and the management of the island conformed to local and national administrative rulings. A large investment was made in the works that were carried out, for which there was some return on the lease of the farm and houses, but unfortunately Christie suffered from mental problems and a series of strokes, which by 1925 had incapacitated him, so that the island was sold by his wife (Blunt, 1968).

THE HARMAN OWNERSHIP

The buyer was Martin Harman, a London businessman who had fallen in love with Lundy when he was a very young man still making his way. He was an enthusiast for the countryside, its birds, flora and fauna, and he delighted in his exceptional island fiefdom (Gade, 1978, passim). He revived and defended energetically Heaven's view of Lundy's independence, and consequently all the mainland authorities were 'dismissed,' with the exceptions of the church and Trinity House, which were immoveable. Thereupon Harman himself undertook responsibility for the postal arrangements, the coastguard, and communications with the mainland.

He set about improving the facilities for his family, and visitors - who were welcomed, provided they shared his concern for the natural environment and, in particular, what he called 'the ancient privileges.' He used the whole of the farmhouse and the north and south annexes to rebuild as one hotel, with the innovations of baths, and a Tavern (Figure 36). There was a tennis court, and a short-lived golf course was opened, traces of which are still to be seen on Ackland's Moor, where enthusiasts still occasionally indulge in some exceedingly rough golf.

Figure 36: The Manor Farm Hotel about 1930. Myrtle Ternstrom collection

The years of the Harman ownership have been described by his agent and friend, F.W. Gade, in his book *My Life on Lundy*. One aspect of Mr Harman's ownership was his introduction of some unusual feral herds, of which Soay sheep, sika deer, goats, and the Lundy ponies remain. Another was his introduction in 1929 of Lundy stamps, which have been enormously successful not only in generating island income, but also in creating a band of island enthusiasts. In 1935 he arranged for an air service from Wrafton, which facilitated his own and visitors' journeys to the island (Figure 37). He was not resident, but the family regarded Lundy as their home, and in the 1930s there was a lively social life with many regular visitors

Figure 37: The airplane in use in the 1930s. Myrtle Ternstrom collection

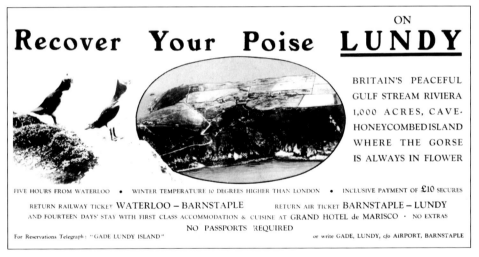

ON
Recover   Your   Poise   LUNDY

BRITAIN'S PEACEFUL
GULF STREAM RIVIERA
1,000 ACRES, CAVE-
HONEYCOMBED ISLAND
WHERE THE GORSE
IS ALWAYS IN FLOWER

FIVE HOURS FROM WATERLOO • WINTER TEMPERATURE 10 DEGREES HIGHER THAN LONDON • INCLUSIVE PAYMENT OF £10 SECURES

RETURN RAILWAY TICKET WATERLOO – BARNSTAPLE   RETURN AIR TICKET BARNSTAPLE – LUNDY

AND FOURTEEN DAYS' STAY WITH FIRST CLASS ACCOMMODATION & CUISINE AT GRAND HOTEL de MARISCO · NO EXTRAS

NO PASSPORTS REQUIRED

For Reservations Telegraph : "GADE LUNDY ISLAND"   or write GADE, LUNDY, c/o AIRPORT, BARNSTAPLE

Figure 38: A poster for Lundy shown in the London Underground in 1939.
Myrtle Ternstrom collection

(Figure 38). Some visitors today see 'Airfield' on the map and expect to see a concrete runway with lights ... and find a bumpy grass strip marked out by white-painted lumps of granite.

Lundy was first seen as a day-trippers' destination in the later years of the nineteenth century as the paddle-steamers from South Wales and the Bristol Channel ports increased in number and frequency (Figure 39, May 1980). By the 1930s, and following the introduction of paid holidays, the number of visitors grew rapidly, and the landing fees contributed very considerably to the island's economy. Since the 1930s Lundy has also been a popular holiday destination for staying visitors - firstly in the hotel and then, increasingly, in cottage accommodation.

Figure 39: Day trippers disembarking from Campbell's steamer. They were ferried to and from the landing stage by launches. *Photo: R. Derek Sach*

The effect that wars have left on Lundy can still be seen in places. The Admiralty built two coastal lookouts at Tibbetts Hill and the northernmost point in 1909 in the expectation of war, and in 1914 the one at Tibbetts Hill was altered to provide accommodation for the coastguards (Figure 40; NT, 1907-08; Lloyd's Archive). The Landmark Trust has removed the superstructures made for the war, and refurbished the building, which still has some of the interior fittings, and is now one of the letting cottages. More than thirty-nine trenches were cut across areas that could have been used as landing grounds at the beginning of the 1939-1945 war and many are still there, as are the last few remains of a Heinkel aircraft that crash-landed on the plateau in 1941 (Figure 41). One other remarkable event for Lundy was the advent of the first mechanical vehicle in 1941, a Fordson tractor, with which the island was supposed to contribute to efforts to grow more food: a pious hope.

Figure 40: The Admiralty watch hut at Tibbetts Hill in 1920. *Photo: H. Jukes (LFS archive)*

Figure 41: Remains of the German Heinkel III aircraft that crashed in 1941.
Photo by kind permission of Roger Davis

The most striking reminder of the war is found in one of the quarries. Mr Harman's elder son, John, lost his life at Kohima during the Burma campaign, and was awarded a posthumous VC for his gallantry. His simple memorial stone is set upon a rock in the VC quarry - a very fitting and moving tribute to a sacrifice by one for whom Lundy had been very dear (Figure 42).

Figure 42: The memorial to John Harman, VC. M. Ternstrom collection

Perhaps Mr Harman's most important contribution, from the point of view of today's meeting, is that he responded very positively to the request by the Devon Birdwatching and Preservation Society to establish an observatory on Lundy. He assisted with the founding of the Lundy Field Society in 1946, gave an initial donation to start it off, and granted the rent-free use of the Old Lighthouse for the Society's members. Conditions were fairly spartan, but the Society appointed a warden, and the members set about establishing bird watching and field studies (Irving *et al.*, 1997).

Although the Society was not able to maintain the Old Light after 1958, its work had expanded to include a wide range of field and marine studies, geology, and archaeology, all of which were published in the Society's Annual Reports and other relevant journals. In celebrating the 60th year of our existence we pay tribute to the founder members and to Mr Harman for their enthusiasm. We also rejoice in the fact that the Society's early studies have led to the establishment of the Marine Nature Reserve, the recognition of Lundy's particular archaeological interest, and its present status as a Site of Special Scientific Interest. In other words, Lundy has arrived on the larger scene, and has 41 scheduled sites and monuments and 14 listed buildings.

After the death of Mr Harman in 1954, and of his son, Albion, in 1968, there was an extensive landslide on the Beach Road, and it was no longer possible for the Harman family to sustain the expenses of the island. It was put up for sale in 1969, and an appeal was launched to raise funds for purchase by the National Trust. This was overtaken by Mr Jack Hayward's generous donation of the whole sum needed. As the National Trust could not acquire the island unless it could ensure that it could be managed financially, the sale was completed when the Landmark Trust, founded by John and Christian Smith, no less generously undertook the lease of the island.

THE NATIONAL TRUST AND THE LANDMARK TRUST

This marked the end of private ownership of Lundy, and 1969 was the start of a period of costly restoration of the buildings and the infrastructure, with the provision of communications by ship, telephone, and, later, helicopters. The Landmark Trust funded the administration of Lundy and the shortfall in its finances for many years until the Lundy Company was formed, as part of the Landmark Trust, and is now responsible for the management of the island. The National Trust has carried out an archaeological survey of Lundy that has listed, mapped and described 170 sites of archaeological and historical interest (Thackray, 1989; NT, 2002).

In the work of restoration and regeneration on Lundy, support has been received from a number of organisations concerned with conservation, heritage, sea fisheries and Countryside Stewardship, that now meet with the island administrator and warden to form a management committee, on which the Lundy Field Society is also represented. Grants for specific projects have been obtained from a number of organisations, of which one remarkable result has been the extension of the Beach Road and the construction of a pier (Figure 43). This is of enormous advantage in the transit of passengers and freight. There are now 24 letting properties available for visitors all the year round, with a popular Tavern catering for all meals, and an excellent shop.

It is seen that changes arose sometimes from the initiatives of Lundy's owners, and sometimes from their neglect. Equally, agents for change have been national concerns, economic forces, social progress, and dramatic increases of interest and concern for historical sites and the natural world.

Despite this modern awareness, the image of Lundy is often presented in the popular media as having to do with pirates and puffins. There are none of the former, and it is very difficult to catch sight of the latter. Also it is remarkable that Lundy has been largely ignored as part of the County of Devon; for example, in

Figure 43: The pier constructed by 1999. The slipway built by Christie is just to the right of it. *Photo: M. Ternstrom*

some topographical and historical works about Devon, Lundy is not even in the index. It is a rich part of Devon's archaeological and environmental heritage, and it is one of the objectives of the Lundy Field Society to bring about an awareness and appreciation of the small treasure that lies off Devon's northern shore.

ACKNOWLEDGEMENTS

I am very grateful to Trinity House for permission to copy and reproduce material from their legal archive, the Minutes of the Boards, and the Engineers' archive.

I also wish to record my thanks to the Heaven family, who have given me generous access to the family papers and photographs, and allowed me to reproduce them in a number of contexts, including the present one.

I have also appreciated very much the kindness of the Mission to Seafarers for permission to use extracts from the unpublished diary of the Reverend John Ashley, kindly transcribed by Miss R. Charles, B.A.

I thank Roger Davis who kindly gave me his photograph of the crashed Heinkel with his permission to reproduce it.

REFERENCES

Ashley: The Revd John Ashley, MS Diary 1841-43, The Mission to Seafarers, London.
BL MSS: British Library MSS, 1822. Peel Papers. Add 40345: 'Free of Tithes, Taxes, Poor Rates, Quit or Chief Rents, or any Outgoing Whatever.'
Blunt, W. 1968. *John Christie of Glyndebourne*, London.
BM: British Museum Maps, 299A, 1809
Bushell, T. 1647: *A Brief Declaration*. British Library, E.401 (3), E.433 (24)
CCR 1243: Calendar of Close Rolls, 1242-47, p.97.
Chanter, J. 1877. *Lundy Island*. London: Cassell.
DRO: Devon Record Office, EM/M/20.
Gade, F.W. 1978. *My Life on Lundy*, p.174. Privately printed.

Gardner, K.S. 1962. 'Preliminary Report on Archaeological Investigations in "The Bulls Paradise" Lundy', 1961. *Annual Report of the Lundy Field Society 1961*, 14, 22-24.

Gardner, K.S. 1971. *An Archaeological Field Guide*. The Landmark Trust.

GPO Archive, Post 30/640, 1893. London. Vols 308, 1886; 336, 1887; 352, 353, 1888.

Grose F., 1776, *Antiquities of England & Wales,* iv, pp. 191-196.

Harman archive.

Heaven Family Papers, Diary, Log, Correspondence, Family History.

Holinshed, R.1586, *Chronicles; the Description of Britaine,* ed. 1807.

Irving, R., Schofield, J., Webster, C., (Eds) 1997. *Island Studies: 50 years of the Lundy Field Society.* Bideford: Lundy Field Society.

The Landmark Trust: Archives, Shottesbrooke. Handbooks, 1968 to date.

Limerick Record office, T22.

Lloyd's Archive, 7778, London.

Luard, H., Ed., 1872, *Chronica Majora of Matthew Paris,* 1240-124.

LFS: Lundy Field Society archive. Annual Reports 1947-.

May, B., 1980, 'The Rise of Ilfracombe as a Seaside Resort in the 19th and 20th Centuries,' *West Country Social & Maritime History*, 137-159, Exeter.

Nat. Arch: National Archives, 1765, Donn, WO 78/5679 1783, ADM 1/ 2307 XC 32991. 1804, Parkyas, MPH/54, WO 1/1110 1892, MT 4/440.

NDRO: North Devon Record Office, B 170 add/37; B 170/192, 1920-25 1840 map (Landmark deposit) Barnstaple.

NT: National Trust Archive, Ref.1287, 1919-1921. London.

NT: National Trust, 2002, *The archaeology and landscape of Lundy: A field guide.*

Steinman Steinman, G. S. 1836. 'Some Account of the Island of Lundy'. *Collectanea Topographica et Genealogica,* iv, 313-330.

Stucley, J., 1983. *Sir Bevill Grenville and His Times.* Chichester: Phillimore.

Ternstrom, M. 1998. 'The Ownership of Lundy by Sir Richard Grenville and his Descendants, 1577-1775,' *Transactions of the Devonshire Association,* cxxx, 65-80.

Ternstrom, M. 1999. *Lundy ... A Study of Factors Affecting the Development of the Island from 1577-1969, with a Gazetteer of Sites & Monuments,* Ph.D. Thesis, University of Gloucestershire.

Ternstrom, M. 2000. 'An Hypothesis for the Origin of the House by the Castle.' *Annual Report of the Lundy Field Society 1999,* 50, 94-101

Ternstrom, M. 2005. 'Granite: A Failed Enterprise on Lundy.' *Transactions of the Devonshire Association,* 137, 193-220.

Thackray, C. 1989. *The National Trust Archaeological Survey.* Unpublished.

THEA: Trinity House Engineers' Archive, 1819, No 1318. 1820, No 1319.

THGM: Trinity House Guildhall Library Minutes, 1819-20. Minutes, 1861-2, Minutes, 1895-7.

Thomas, S.1959. *The Nightingale Scandal,* Bideford. Reprinted 2001.

Veale, E. 1957. 'The Rabbit in England,' *Agricultural History Review,* v, pt 2, 85-90.

For the later material in this account I have to a considerable extent used my own notes, which date from 1952.

ARCHAEOLOGY AND HISTORY: DISCUSSION

(Initials: HQ=Henrietta Quinnell, MT=Myrtle Ternstrom, SB=Shirley Blaylock, JH=John Hedger, KG=Keith Gardner, AW=Ann Westcott, Q=Unknown participant)

JH: Has the history of the island i.e. past land use affected the vegetation?

MT: There were fires in Benson's time (mid 1700s) and extensive fires at the North End in 1933. There are nineteenth century records of vegetation, but we need pollination studies.

SB: Obviously depending on the climate the vegetation comes back with the passage of time. The greatest impact of land use is at the South End where farming was prevalent.

JH: There appears to be something odd about Middle Park?

SB: Middle Park (Tibbetts area) was ploughed more recently. On the east side previous field boundaries survive; there is evidence that there was cultivation there from the Bronze Age. The North End has a more 'natural' landscape in that there was no medieval cultivation.

KG: Some pollen analysis has been attempted in the Middle Park area and results showed a reversion from grassland to heathland.

KG: The aceramic Iron Age - was it abandoned completely?

HQ: The whole of Wales and Ireland did not use pottery through from the late Bronze Age, the Iron Age and through the Roman Period, except for obviously imported traditions. This lack of ceramics was evident to a considerable extent in Devon. Lundy probably reflects the practices in these areas with which it was in contact. However it should be stressed that many communities did not use pottery in the Iron Age, but the lack of artefacts does not mean that sites did not exist e.g. people probably used wooden or stone tools that have not survived.

KG: In the Early Christian Period, there were some imported Mediterranean wares. A sherd has been found in the Brickfields.

HQ: The lack of ceramics in the post-Roman Period is very well established throughout the South West. There is a little local production in Somerset, but indigenous pottery was only made in western Cornwall.

Q: A layer of clay under the church was mentioned?

MT: A 14ft bed of clay was discovered under the church when the foundations were dug. We have little information and more geological investigation is needed,

Q: What were the sea-level changes in the Stone and Bronze Ages? Was the island larger in size than it is now?

HQ: The study of sea level change is very complex and outside the scope of my expertise. At the end of the Ice Age sea level was lower than it is today, probably about 100m. Lundy in the prehistoric period was probably broadly similar in extent to what it is today.

Q: What animals, mammals, were hunted with the early arrow heads found?
HQ: There is no specific evidence of mammal populations present at that time. Bones do not survive in Lundy's acid soil nor are there middens where limpet shells counteract the effects of acid soil. Hunter-gatherers could have hunted red deer that can reach islands, but there is no evidence of their presence on Lundy. Obviously seals and sea-birds were present at that time and were probably hunted.

AW: Is it possible to have a good pollen analysis done on the island?
SB: Pollen samples have been taken during an archaeological survey by English Heritage. Funding is needed and also a radiocarbon dating programme.

INTRODUCTION TO THE MARINE AND FRESHWATER HABITATS OF LUNDY

by

KEITH HISCOCK

Marine Biological Association, Citadel Hill, Plymouth, PL1 2PB
e-mail: khis@mba.ac.uk

ABSTRACT

The marine and freshwater habitats of Lundy are populated by very different species and communities and have a contrasting history of habitat creation and colonisation. Marine habitats are largely natural and are very varied, particularly reflecting the wide range of conditions of wave and tidal stream exposure around the island. The variety of freshwater habitats is surprising for such a small island, including ponds, cisterns and reservoirs as well as the streams. Both marine and fresh waters have great fascination and importance for their natural history and both are special features of Lundy.

Keywords: *Lundy, ponds, streams, freshwater habitats, marine biology, marine habitats, conservation*

INTRODUCTION

Both salt and fresh water have always been important to Lundy. The sea, before its special wildlife features were revealed, for its fish and shellfish; the ponds and streams because access to freshwater was essential for human survival on the island. Both have been affected by human activities but both have an enormous fascination and value for the wildlife that they support.

The major groups of plants and animals that populate the shore and seabed around Lundy are very different to those of freshwaters on the island. In the sea, algae are the plant species colonising the seabed whilst, in freshwaters, it is flowering plants (angiosperms) that are the attached species. In the sea, sponges, sea anemones and their relatives (Cnidaria), polychaete worms, crustaceans, molluscs, bryozoans (sea mats), echinoderms (sea urchins, starfish and their relatives) and ascidians (sea squirts) are dominant whilst, in freshwaters, the variety of major animal groups is much smaller with crustaceans and insects dominant and a few flatworm, annelid worms, arachnid (spiders) and mollusc species also present.

My first visit to stay on Lundy was in 1967 and I dived there for the first time to study marine wildlife in 1969. Since then, I have returned to the island in most years and, in the 1970s and '80s, brought many colleagues to help in documenting the marine wildlife - admittedly therefore being on the sea rather than the land, and anywhere near Lundy's fresh waters.

MARINE HABITATS

Marine habitats and the communities of species that populate them are shaped by wave action, tidal flow velocity, the underlying geology including geomorphology and sediment types, depth to the seabed, 'water quality' (meaning especially salinity, turbidity, nutrients) and the prevailing currents that bring water masses and larvae. Studying maps and charts of Lundy as well as the 'Coastal Pilot' makes it clear that there are going to be a wide range of habitats there. Both granite and slate rocks slope steeply into the subtidal and extend to considerable depths in places. The shore and seabed are exposed to extremely strong wave action on the west coast but are much less exposed on the east coast, and there are very strong tidal flows off the north and south ends of the island compared to 'weak and intermittent' currents off parts of the west and east coast. There are extensive sediments off the east coast. The island stands at the meeting point of clear oceanic waters to the west and turbid Bristol Channel waters to the east suggesting influence from both.

Despite the difficulties of getting there, Lundy has attracted the attentions of marine biologists for over 150 years. The earliest recorded studies near to Lundy are noted in the work of Forbes (1851) who took dredge samples off the east coast of the island in 1848. The first descriptions of the seashore wildlife on Lundy are those published in 1853 by the foremost Victorian marine naturalist and writer, P.H. Gosse, in the *Home Friend* (reproduced later in Gosse 1865). However, his descriptions are unenthusiastic, reveal nothing unusual and draw attention to the very few species found on the granite shores. There are further brief references to Lundy in the literature of other Victorian naturalists. Tugwell found the shores rich collecting grounds and cites the success of a collecting party who (with the help of 'an able-bodied man with a crowbar') returned from Lundy in 1851 'laden with all imaginable and unimaginable spoils' (Tugwell, 1856). However, Lundy never achieved the popularity of the nearby North Devon coast amongst Victorian sea-shore naturalists and significant published studies of the marine life of the island did not appear until the 1930s.

Each summer between 1934 and 1937, G.F. Tregelles visited Lundy to collect seaweeds. His records are summarised in Tregelles (1937) and are incorporated into the *Ilfracombe fauna and flora* (Tregelles *et al.*, 1946) and the *Flora of Devon* (Anonymous, 1952).

The first systematic studies of marine ecology at Lundy were undertaken by Professor L.A. Harvey and Mrs C.C. Harvey together with students of Exeter University in the late 1940s and early 1950s (Anonymous, 1949; Harvey, 1951; Harvey, 1952). These studies again emphasised the richness of the slate shores especially when compared with the relatively impoverished fauna on the granite shores. A later study (Hawkins & Hiscock, 1983) suggested that impoverishment in intertidal mollusc species was due to the isolation of Lundy from mainland sources of larvae.

When marine biologists started to use diving equipment to explore underwater around Lundy at the end of the 1960s, they discovered rich and diverse communities and many rare species. These finds led to a wide range of studies being undertaken, both underwater and on the shore, in the late 1960s through to the mid-1980s. The flora and fauna were catalogued and ecological studies resulted in a detailed knowledge of the inshore marine biology of the island, contributing significantly to the understanding of subtidal marine ecology in Britain. A summary description of the marine ecology of Lundy is given in Hiscock (1997). More recently, particularly as Lundy became Britain's first marine nature reserve, surveillance studies have revealed the great longevity of many species and their likely irreplaceability if damaged. The wide range of studies undertaken are catalogued in Hiscock (1997).

Although the great interest and value of Lundy's marine life is in natural habitats, there are habitats that result from human activities or which have been affected by human activities. The wrecks around Lundy are mostly flattened and now merge with the surrounding seabed except that the M.V. *Robert* (which sank off the east coast in 1975) is intact and has a community of species not found on natural substrata. The jetty, built in 1998, has particular communities of species on the pilings. However, the richness of the rockpools that once existed in the Landing Bay is now severely degraded by the spoil from the beach road excavations in the late 1980s.

Although the 'connectedness' of the sea means that larvae, spores and migratory species can readily colonise the island from afar, there remain mysteries as to how certain species with short-lived larvae reached the island. Our knowledge of reproductive biology of species is far from complete but some such as sea fans (*Eunicella verrucosa*), the sunset cup coral (*Leptopsammia pruvoti*) and most likely many of the sponges have very short-lived larvae and now only recruit locally - so how did they get to the island in the first place?

The richness and composition of marine life around Lundy is not static. The profusion of colourful Mediterranean-Atlantic species, especially sponges, corals and anemones, may have reached a 'high point' in the '70s or '80s but has been in decline since the mid '80s (Hiscock, 2003) for no clear reason except that there may be some variability that is so long-term we do not yet recognise it. In 2001 and 2002, the sea fan (*Eunicella verrucosa*) population was decimated by a bacterial infection, now passed. Non-native species have appeared - most conspicuously japweed (*Sargassum muticum*) in the Landing Bay and climate warming is encouraging some previously sparse southern species to thrive (notably the toothed topshell *Osilinus lineatus* in the Devil's Kitchen).

Lundy was established as a voluntary marine nature reserve in 1972 (Hiscock *et al.*, 1973) and as a statutory marine nature reserve in 1986. The area around Lundy is a Special Area of Conservation established under the EC Habitats Directive and has the only No-Take Zone, established by fisheries bye-law, to protect wildlife in the U.K. Now, much of the marine biological research is focussed on monitoring the effectiveness of the various conservation measures.

FRESHWATER HABITATS

Freshwater habitats include streams and standing waters: ponds, cisterns and reservoirs. On Lundy, the soil is acidic and therefore streams and ponds tend towards acidic rather than alkaline and, because dissolved mineral levels are low, are 'soft' rather than 'hard'. Both of those factors will affect the communities of plants and animals that develop in freshwaters but colonisation of those waters by plants and animals on an isolated island must be more problematic than for the sea. Plants and animals, including their propagules, may be brought to the island by birds, and humans have doubtless introduced some. Long (1994) has noted how Lundy streams are impoverished compared to similar streams on the mainland. However, George (1997) also observes that there are usually fewer parasites, predators and competing species present in such isolated locations.

Lundy is a small island and most of the rain that falls is likely to run away to the sea without ponding. However, the largest standing water body, Pondsbury, is natural, albeit modified by damming and dredging. Many of the standing freshwater habitats were created by human activity to provide a regular supply of water. The older examples of such habitats are open and colonised by plants and animals. But many of the freshwater habitats are in danger of drying-up during extended periods of dry weather, placing significant stress on the component flora and fauna.

Lundy does not have a long history of study of freshwater habitats. The earliest recorded observations of the freshwater flora and fauna are those of Morgan (1948) who studied the streams and Fraser-Bastow (1950) who studied diatom algae. It seemed likely that the field courses from Exeter University run by one of the LFS founders, Professor L. A. Harvey, would have sampled and documented pond life, In fact, although marine records are very full from those courses, there is nothing in the original field records to suggest what could be found in freshwater.

The variety of freshwater habitats is catalogued by Langham in the *Annual Report* of the LFS for 1968 (Langham, 1969). In that *Annual Report*, location, size and dominant vegetation was recorded. The first detailed records of freshwaters was that undertaken by Jennifer George, Brenda McHardy (Stone) and others. Information collected up to 1996 was included in the review by George (1997).

George (1997) concludes that the isolation of Lundy is not a major limiting factor for the freshwater fauna and flora but that drying-up of habitats during drought is an important environmental factor. Whilst it seems that, in the 27 years that Jennifer George and her colleagues have studied freshwaters, there has been a high degree of constancy in the fauna and flora present (see George, this volume), long-term variability is uncertain and there will doubtless be surprises in the future.

Although most of the island is scheduled as a Site of Special Scientific Interest, there is no mention of freshwaters in the citation.

Plates 1-25 on pages 68-80 show the diversity of the marine life occurring in the Lundy Marine Nature Reserve.

REFERENCES

Anonymous. 1949. Marine ecology. *Annual Report of the Lundy Field Society 1948*, 2, 28-33.

Anonymous. 1952. *Flora of Devon. Volume II, Part I. The marine algae*. Torquay.

Forbes, E. 1851. *Annual Report* on the investigation of British marine zoology by means of the dredge. Part 1. The infralittoral distribution of marine invertebrata on the southern, western and northern coasts of Great Britain. *Annual Report of the British Association for the Advancement of Science*, 20, 192-263.

Fraser-Bastow, R. 1950. Freshwater diatom flora. *Annual Report of the Lundy Field Society 1949*, 3, 32-41.

George, J.J. 1997. The freshwater habitats of Lundy. In: *Island studies: 50 years of the Lundy Field Society*, ed. by R.A. Irving, A.J. Schofield, & C.J. Webster, 149-164. Bideford: Lundy Field Society.

Gosse, P.H., 1865. *Land and sea*. London, James Nisbet.

Harvey, L.A. 1951. The granite shores of Lundy. *Annual Report of the Lundy Field Society 1950*, 4, 34-40.

Harvey, L.A. 1952. The slate shores of Lundy. *Annual Report of the Lundy Field Society 1951*, 5, 25-33.

Hawkins, S.J. & Hiscock, K. 1983. Anomalies in the abundance of common eulittoral gastropods with planktonic larvae on Lundy Island, Bristol Channel. *Journal of Molluscan Studies*, 49, 86-88.

Hiscock, K., Grainger, I.G., Lamerton, J.F., Dawkins, H.C. & Langham, A.F. 1973. A policy for the management of the shore and seabed around Lundy. *Annual Report of the Lundy Field Society 1972*, 23, 39-45.

Hiscock, K. 1997. Marine biological research at Lundy. In: *Island studies: 50 years of the Lundy Field Society*, ed. by R.A. Irving, A.J. Schofield, & C.J. Webster, 165-183. Bideford: Lundy Field Society.

Hiscock, K. 2003. Changes in the marine life of Lundy. *Annual Report of the Lundy Field Society 2002*, 52, 86-95.

Langham, A.F. 1969. Water courses and reservoirs on Lundy. *Annual Report of the Lundy Field Society 1968*, 19, 36-39.

Long, P.S. 1994. A study into the macroinvertebrate fauna and water quality of Lundy Island's lotic environment. *Annual Report of the Lundy Field Society 1993*, 44, 59-72.

Morgan, H.G. 1948. Aquatic habitats. *Annual Report of the Lundy Field Society 1947*, 1, 12-13.

Tregelles, G.F. 1937. An introduction to the seaweeds of Lundy. *Annual Report of the Transactions of the Devonshire Association for the Advancement of Science*, 69, 359-363.

Tregelles, G.F., Palmer, M.G. & Brokenshaw, E.A. 1946. Seaweeds of the Ilfracombe district. *In: The fauna and flora of the Ilfracombe district of North Devon*, ed. by M.G. Palmer. Townsend & Sons, Exeter, for the Ilfracombe Field Club.

Tugwell, G. 1856. *A manual of the sea-anemones commonly found on the English coast*. London: Van Voorst.

Plate 1: Boulders at Ladies Beach provide a habitat for a rich variety of algae and animals, especially under the boulders. *(Photo: Keith Hiscock)*

Plate 2: The anemone, *Actinia fragacea*, in a rock pool on Divers Beach. *(Photo: Keith Hiscock)*

Plate 3: The scarlet and gold star coral, *Balanophyllia regia*, in a gully at Mouse Island. *(Photo: David George)*

Plate 4: *Lepadogaster lepadogaster*, the shore clingfish or Cornish sucker, under boulders at the jetty. *(Photo: Keith Hiscock)*

Plate 5: The black brittle star, *Ophiocomina nigra*, north of Rat Island.
(Photo: Keith Hiscock)

Plate 6: The seven-armed starfish, *Luidia ciliaris*, north of Rat Island.
(Photo: Keith Hiscock)

Plate 7: On the bridge of the M.V. *Robert* wreck, showing the plumose anemone, *Metridium senile*, the sea fir, *Nemertesia antennina*, and a red alga. *(Photo: Keith Hiscock)*

Plate 8: A conger eel, *Conger conger*, framed by plumose anemones on the M.V. *Robert. (Photo: Keith Hiscock)*

Plate 9: Undisturbed sediment on the east coast showing the spiny starfish, *Marthasterias glacialis*, and the anemone, *Mesacmaea mitchelli*. *(Photo: Keith Hiscock)*

Plate 10: The branching yellow axinellid sponge, *Axinella polypoides*, on a boulder slope off the Quarries, east coast. *(Photo: David George)*

Plate 11: The edible crab, *Cancer pagurus*, 'safe' in the No-Take Zone on the east coast. *(Photo: David George)*

Plate 12: The blue-spot sea slug, *Greilada elegans*, which was formerly common around Lundy but which has not been seen since the mid-1980s. *(Photo: Keith Hiscock)*

Plate 13: Slope at the Knoll Pins showing the sea-fan, *Eunicella verrucosa*, and the red sea-fingers, *Alcyonium glomeratum. (Photo: Keith Hiscock)*

Plate 14: Close-up of *Alcyonium glomeratum* at the Knoll Pins.
(Photo: Keith Hiscock)

Plate 15: The colonial yellow cluster anemone, *Parazoanthus axinellae*, on the Knoll Pins. *(Photo: David George)*

Plate 16: The crab, *Inachus phylangium*, in the snakelocks anemone, *Anemonia viridis*, on the Knoll Pins. *(Photo: Keith Hiscock)*

Plate 17: Scene at Gannets Rock showing the spider crab, *Maia squinado*, the sea-urchin, *Echinus esculentus*, ross, *Pentapora foliacea*, and the yellow boring sponge, *Cliona celata*. *(Photo: Keith Hiscock)*

Plate 18: The jewel anemone, *Corynactis viridis*, on a vertical rock face at Gannets Rock. Asexual reproduction by the parent polyp produces clumps which are all of one colour. *(Photo: David George)*

Plate 19: Underwater life at the Hen and Chickens, north coast, showing beds of the hydroid, *Tubularia indivisa*, the boring sponge, *Cliona celata*, and jewel anemones, *Corynactis viridis*. *(Photo: Keith Hiscock)*

Plate 20: The lobster, *Homarus gammarus*, emerging from a rock fissure at Jenny's Cove, west coast. *(Photo: Keith Hiscock)*

Plate 21: The crawfish or spiny lobster, *Palinurus elephas*, showing its heavily armoured body and long antennae, emerging from under a boulder at The Rattles Anchorage, south coast. *(Photo: David George)*

Plate 22: An Atlantic grey seal, *Halichoerus grypus*, swimming in Gannets Bay, east coast. *(Photo: Miles Hoskin)*

Plate 23: The jellyfish, *Cyanea lamarckii*, with its long stinging tentacles, in the Landing Bay. *(Photo: Keith Hiscock)*

Plate 24: The compass jellyfish, *Chrysaora hysoscella*, with its distinctive markings travelling along the east coast. *(Photo: David George)*

Plate 25: Divers monitoring scallop populations off the east coast. The densities and sizes of the scallop, *Pecten maximus*, that occur within the width of the pipe as it is taken along a set distance, are measured. This procedure is part of the monitoring of the No-Take Zone. *(Photo: Keith Hiscock)*

LUNDY'S MARINE LIFE - A BALANCING ACT OF PROTECTING AND PROMOTING

by

ROBERT IRVING

Combe Lodge, Bampton, Devon, EX16 9LB
e-mail: Robert@sea-scope.co.uk

ABSTRACT

The waters around Lundy exhibit a great wealth of marine habitats and wildlife rarely seen in such a small area. A statutory Marine Nature Reserve was established around the island in 1986, and since then the same area has also become a Special Area of Conservation. In 2003 part of the area was also designated a No-Take Zone. The purpose and function of these designations is explained and an assessment given of what effect they may have had on Lundy's marine life. The need for protecting Lundy's near-shore seabed is examined. It is concluded that the protection offered is adequate but could be improved in certain areas. A review is given of the various means by which the area has been, and is being, promoted.

Keywords: *Lundy, Marine Nature Reserve, Special Area of Conservation, marine life, protection, management, No-Take Zone*

INTRODUCTION

Although it is only a small island, and consequently in the minds of many people fairly insignificant, within the sphere of marine nature conservation Lundy is one of the top sites within British waters. Indeed, Natural England[1] has described the waters around the island as being 'the jewel in the crown' of their marine nature conservation policy (English Nature, 1993). So what is it that makes Lundy's marine life and seabed habitats be regarded so highly, and what measures are in place to ensure that these special features are maintained for future generations? Has the creation of a marine nature reserve around the island helped or hindered the marine life and seabed habitats? Is it feasible to 'manage' a marine nature reserve in a similar way to managing a nature reserve on land? Should one overtly publicise these natural riches beneath the waves, or should one keep quiet about them, in order to minimise possible disturbance? Should commercial exploitation of a resource within the area (such as fishing) be allowed to continue whilst at the same time ensuring that adequate protection is afforded to delicate habitats and species?

---

[1] Natural England is the Government funded body whose purpose is to promote the conservation of England's wildlife and natural features. This includes marine habitats and species. Until October 2006 it was known as English Nature, but after its amalgamation with the Countryside Commission and the Rural Development Service it has become Natural England.

This paper seeks to address these and other questions by reviewing what is known about the island's marine life and habitats to date; by assessing the need for protecting this natural resource and how such protection has been put in place; and by reviewing the means by which the area has been promoted since becoming England's first and only (to date) statutory marine nature reserve.

WHAT MAKES THE MARINE LIFE AROUND THE ISLAND 'SPECIAL'?

There are several reasons why Lundy's marine life is regarded as being of particular note. Within the SCUBA diving community, the island is well-known for its spectacular underwater scenery, its clear waters (especially if visited after a spell of calm weather) and its colourful marine life. There is no doubt that, for such a small area, a remarkable variety of marine habitats are present around the island, with each displaying its own characteristic wildlife. The niches available to would-be colonisers are further enhanced by the range of environmental conditions which the island experiences. Indeed, Natural England (previously English Nature) proudly boasts on its website *(http://www.english-nature.org.uk)* that 'Lundy has the finest diversity of any marine site in the U.K.'

As well as escaping mainland sources of pollution, being an offshore island has other advantages. Simply by being an island, Lundy has an exposed side and a sheltered side to it. The prevailing wind direction is from the south-west, leading to the west and south coasts being exposed to the full force of Atlantic gales, while the east coast remains relatively sheltered. Not only is this reflected in the shore biota (the west coast shores are dominated by barnacles and limpets with very few seaweeds apparent; the east coast shores, by contrast, have a far more diverse biota with lush seaweed growths), but also in the seabed types. The west coast is dominated by huge 'slabs' of granite bedrock scoured by sand trapped in the base of gullies, whilst the sheltered east coast has large areas close inshore of mud or muddy gravel. Off the south-east coast the bedrock is of slate, which fragments into smooth, flat pebbles forming mobile areas that are frequently colonised by beds of brittlestars.

With its north-south orientation, the island acts as a breakwater across the flow of the tides up and down the Bristol Channel. The tidal range at Lundy is almost 8 m (on spring tides), leading to strong currents being experienced around the north-west, south-west and south-east corners of the island in particular. Lying 11 miles from the nearest point of the mainland, Lundy is also on the border between a coastal, relatively murky, body of water and an oceanic, relatively clear, body of water. This situation also contributes to the existence of conditions which are conducive for a wide range of species to flourish. Bedrock reefs extend to well over 1 km offshore from the west coast and, unusually, deep water (30-40 m) is found relatively close to the island (particularly off the north and north-east coasts). Steeply-sloping, vertical and overhanging underwater cliffs are present here, typically covered by dense growths of sessile marine invertebrates and providing impressive underwater scenery for divers. Indeed, the variety of habitats and the associated species on Lundy's reefs is outstanding and includes, for example, over 300 species of seaweeds and many rare or unusual invertebrate species. The diversity of habitats is further enhanced by the large number of shipwrecks which are present, as well as man-made structures such as the jetty pilings in the Landing Bay.

As a result of its geographical position, Lundy acts as an outpost for several species whose centres of distribution lie much further to the south, often as far south as the Mediterranean. The reason for the occurrence of these species at Lundy is the influence of the north-flowing Lusitanian current emanating from the Mediterranean. This current is slightly warmer than the Atlantic water around it and, from time to time, it may bring with it larval forms, a few of which may be able to survive the slightly cooler northern waters. However, the populations of certain of these Mediterranean-Atlantic species, such as the sunset cup coral *Leptopsammia pruvoti*, are dwindling in size and now exist as isolated, 'relict' populations (see Box 3 below). Several of these Mediterranean-Atlantic species are rare and, being at the extreme edge of their range, are particularly susceptible to changes in their environment. As a result, several have been the subjects of monitoring studies. These pioneering studies, undertaken from 1984 to 1991, have confirmed the slow growth and longevity of many of the species of high conservation interest, such as the sea fan *Eunicella verrucosa*, the sunset cup coral *Leptopsammia pruvoti* and various species of erect sponges (Fowler & Pilley, 1992).

A summary of the wide range of Lundy's intertidal and subtidal habitats and wildlife, and the various studies that have been undertaken on them over the years, is given by Hiscock (1997).

THE ESTABLISHMENT OF THE MNR AND THE SAC

Whilst SCUBA diving as a recreational sport was in its infancy during the 1960s, a growing number of enthusiasts were keen to be the first to visit 'un-dived' locations around the British coastline. Lundy quickly became recognised as a sought-after place to visit, though getting to and from the island was not at all straightforward. A few of those that did make it sought to collect souvenirs of their underwater exploits, and would bring up items such as sea urchins and sea fans at the end of their dives as mementoes. Spearfishing was also popular (frequently leading to the largest fish within a population being removed), as was hunting for lobsters, crawfish and scallops (with the same result). By the end of the 1960s, diving biologists recognised that Lundy had an exceptionally rich variety of marine life which was threatened by certain of the aforementioned practices. Around the same time there was also a growing worldwide movement to establish marine parks and reserves, and Lundy seemed an obvious candidate. However, it was not simply a matter of announcing that a marine reserve had been set up. It took a great deal of persuasion to convince the many interested parties of the need to protect the island's marine habitats and species, and a long and arduous course has had to be followed to reach the position we are now at (as set out in Table 1).

Following initial recognition of the scientific importance of the island's marine habitats and wildlife, a voluntary marine nature reserve was set up in 1972. This worked well to begin with, but as time went by it was apparent that tougher measures were needed to protect the area. There was a constant threat that someone might use dredging gear to take scallops from off the east coast and at the same time cause irreparable damage to the communities of high nature conservation interest there. With the Wildlife and Countryside Act becoming law in 1981, it became possible to establish statutory Marine Nature Reserves (MNRs) which were accompanied by byelaws. However, there was a considerable amount of opposition to the proposals, particularly from fishermen who could see the designation being 'the thin edge of the wedge' and that many more MNRs would sprout up elsewhere in no time. It took several years to allay these fears and eventually a statutory MNR around Lundy was declared in November 1986. A more comprehensive history of Lundy's marine nature reserve (up until 1996) is given by Irving & Gilliland (1997).

Lundy's Special Area of Conservation (SAC) status came about after the adoption of the Habitats Directive into U.K. law in 1994. This Directive, correctly referred to as 'Council Directive 92/43/EEC on the Conservation of natural habitats and wild fauna and flora', requires EU Member States to create a network of protected wildlife areas across the European Union, collectively known as Natura 2000 sites. These sites include both SACs and Special Protection Areas (SPAs), the latter being sites designated for their bird life interest. The U.K. statutory provisions applying to Natura 2000 sites are contained in the Conservation (Natural Habitats &c) Regulations 1994. Initially, SACs (both terrestrial and marine) were notified as 'candidate' sites (cSACs), with formal designation as SACs not taking place until April 2005.

Table 1: Major steps in the protection of Lundy's near-shore waters

DATE	DESIGNATION	NOTES
1972	Formal recognition of a voluntary marine nature reserve (VMNR) around the island, the first of its kind in the country.	Covered foreshore and sea bed from High Water Mark to 1 km offshore. Sufficiently large to include habitats and species of high scientific interest, yet small enough to monitor activities within it. Excluded main fishing banks. *Ref.* Hiscock *et al.* (1973).
1979	'Gentleman's agreement' between fishermen and conservationists to observe a ban on dredging/bottom trawling west of a line between the Knoll Pins and Surf Point.	Brought about to protect, in particular, the population of burrow-dwelling red band fish and other communities present in soft sediment areas. *Ref.* Hiscock (1983).
1985	Formation of the Lundy Marine Consultation Group (re-named in 1994 the Lundy Marine Nature Reserve Advisory Group)	One of the main aims of the Group was to provide a forum for exchanging views on present and proposed activities around Lundy. *Ref.* Cole (1986); & e.g. Irving (2003).

1986	Designation of a statutory Marine Nature Reserve (MNR) around the island, the first such reserve in British waters (Fig.1).	Designated under the Wildlife and Countryside Act 1981 on 21 November 1986, following a 3-month period of notification and 4 years of consultation! Included new DSFC byelaws restricting certain fishing practices. *Ref.* Nature Conservancy Council (1987).
1990	Designation of two of the island's many wrecks, the *Iona II* and the 'Gull Rock site', as protected sites.	Designated under the Protection of Wrecks Act 1973. The *Iona II* was a paddlesteamer built as a fast ferry for the Clyde in 1863 but sank a year later on her way to America. No wreck has been found at the Gull Rock site, but several stone shot and other artefacts dating from the sixteenth century have been found. *Ref.* Robertson & Heath (1997).
1994	Publication of a Management Plan covering the MNR and the (terrestrial) SSSI.	One of the aims of the Plan was to 'establish an effective structure for overseeing the management of the reserve'. A Management Group was formed from the statutory bodies involved in the management of the MNR. *Ref.* English Nature (1994).
1994	Launch of the Zoning Scheme, allocating different zones for different activities within the MNR.	A 'useful tool' pioneered in marine reserves abroad for summarising byelaws and other regulations in an easy-to-understand visual way. *Ref.* English Nature (1995).
1996	Notification by the Department of the Environment as a 'candidate' Special Area of Conservation (cSAC), and in 2005 as an officially recognised SAC by law.	Notified under the EC Habitats Directive (1992) for certain of its marine habitats and species (rocky reefs, shallow sandbanks, sea caves and grey seals).
2003	Designation of the No-Take Zone off the island's east coast - the first such statutory area in the country to ban all forms of fishing within it.	Primarily established to protect vulnerable habitats and species of conservation importance off the east coast, by means of Devon Sea Fisheries Committee byelaws. Popularly viewed as a means of enhancing numbers and sizes of commercially exploitable species.
2005	Formal designation of the Special Area of Conservation (SAC).	Designateed by the Secretary of State for Environment, Food and Rural Affairs on 1 April 2005.

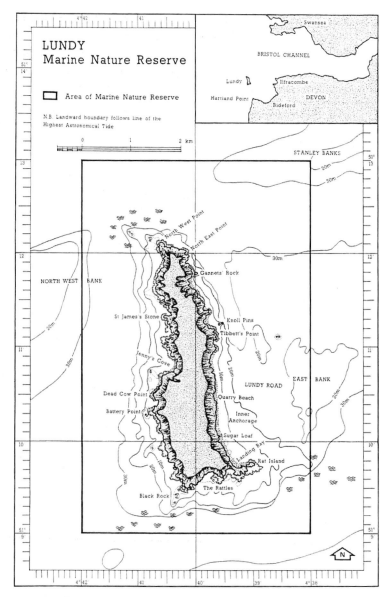

Figure 1: The 'box' boundary to the statutory MNR, which came into force on 21 November 1986 (Nature Conservancy Council 1986). Keeping the seaward boundary of the MNR to straight N/S and E/W lines between four points was done on purpose to simplify the identification of the area for both legal representations (byelaws) and for visitors to/users of the area. Note that the boundary to the SAC is exactly the same.

HOW DO THE MNR AND THE SAC DIFFER?

Both the MNR and the SAC share the same boundaries (Figure 1) - the 'inner' boundary being the island's high water mark and the 'outer' one following the four sides of a rectangle between 1-2 m offshore. For all other purposes, the two designations are quite separate, though they do share the same overall conservation goals. One of the major benefits of the MNR has been the provision of an on-site full-time Warden, which SAC status alone would not have merited (although the site would have had a dedicated SAC Officer). One other point of difference is that the SAC status is based on certain named features of the site (see Table 2), whereas the MNR status covers all the habitats and wildlife within its boundary. Lundy was selected as an SAC on account of its 'outstanding representation of reef habitats in south-west England' *(http://www.jncc.gov.uk)*. Designation as a SAC also brings with it international recognition for Lundy's marine habitats and wildlife.

Table 2: Named features for which the Lundy SAC was designated

FEATURE		
Reefs	An Annex I habitat - the primary reason for the selection of Lundy. The reef extent, water clarity, water temperature and salinity are all attributes to be monitored.	
	SUB-FEATURE	ATTRIBUTES TO BE MONITORED
	Rocky shore communities	Distribution of characteristic range of biotopes; species composition of rockpool communities; characteristic species: Devonshire cup coral *Caryophyllia smithii* and scarlet and gold star coral *Balanophyllia regia*.
	Kelp forest communities	Distribution and range of kelp biotopes; algal species composition.
	Subtidal vertical and overhanging circalittoral rock communities	Species composition of characteristic biotopes.
	Subtidal bedrock and stable boulder communities	Distribution and range of circalittoral biotopes; distribution and extent of sea fan *Eunicella*-characterised reef; species composition of characteristic biotope (MCR.ErSEun); characteristic species - density and quality of sea fans *Eunicella verrucosa*; species composition of sponge-dominated biotope (MCR.ErSPbolSH).
FEATURE	COMMENT	
Sandbanks which are slightly covered by seawater all the time	Annex I habitat present as a qualifying feature, but not a primary reason for selection of this site. The two (of the four) main sub-types which occur at Lundy are: gravelly and clean sands; and muddy sands.	

Submerged or partially submerged sea caves	Annex I habitat present as a qualifying feature, but not a primary reason for selection of this site. According to Hiscock (1982), there are 37 known intertidal caves on Lundy - though this figure may be an under-estimate. Many of the caves extend for tens of metres into the island. There are also a number of subtidal caves.
Grey seal *Halichoerus grypus*	Annex II species present as a qualifying feature, but not a primary reason for selection of this site. Lundy is an important pupping site for grey seals, with approximately 10% of pups (about 20 individuals) born in the south-west region annually (Duck, 1996). Numbers of adults vary but are in the region of 70-120 individuals (Irving, 2005).

Monitoring studies of certain intertidal and subtidal species and communities of particular interest were initiated in 1984, although these studies were not a requirement of the MNR. However, there is now a legal obligation for Natural England to report on the overall condition of the listed 'features' of the SAC (see Table 2) once every six years. The report, submitted to Brussels, is required to state whether the feature in question is being maintained in a 'favourable condition' or not. There are various targets which need to be met before the feature's condition can be said to be favourable. Many of these targets require monitoring work to be undertaken in order to provide the information on which to base the judgement. For the first reporting round (submitted in 2006), monitoring of the intertidal and subtidal reef features was undertaken during the summers of 2003 and 2004.

WHY DOES THE AREA NEED PROTECTING?

Lundy is offered a certain degree of protection by its geographical position alone, thereby avoiding much of the human-generated disturbance (recreational, commercial or industrial) that would affect the area if it were adjacent to the mainland coast. However, there are certain activities which are known to have an impact on the seabed around the island and which are likely to harm the wildlife interest.

The most obvious of these are certain fishing activities, particularly those which are known to disturb the seabed, such as bottom trawling, scallop dredging or tangle netting. The extensive muddy sediment area off the east coast hosts an array of rare and vulnerable species, such as the burrowing anemones *Mesacmaea mitchellii* and *Halcampoides elongatus*. It is this habitat in particular which would suffer as a result of such destructive practices. Recovery of such areas from the impact of towed fishing gear can take several years, and even then the community which develops is likely to show differences to the original community for an even longer length of time. Potting for crustaceans has far less of an impact and consequently this activity has been allowed to continue within most of the MNR/SAC, though it too has now been banned from within the No-Take Zone off the island's east coast.

Commercial fishing is not the only activity which may harm the marine life. SCUBA diving too may result in damage to certain habitats and/or species. Consequently, divers to the MNR/SAC are asked to abide by a Code of Conduct which instructs them (i) to demonstrate good buoyancy control (contact with the seabed should be avoided wherever possible); (ii) to avoid careless finning (inadvertent fin strokes can damage erect species such as certain sponges, sea fans and cup corals); (iii) to avoid disturbing the marine life by direct contact; and (iv) to remember that exhaled air bubbles can lodge in subtidal caves and kill the marine life there.

The Code of Conduct has always requested that anglers return to the sea any territorial fish they may catch. These include all of the five wrasse species (ballan, cuckoo, goldsinny, rock cook and corkwing) and conger eels. These species are long-lived and are likely to remain in or return to the same area over many years. Discarded angling hooks, weights and line can also create hazards to marine wildlife (and also to divers), though this is a relatively minor problem. Angling is now prohibited anywhere within the No-Take Zone.

Other threats to the area are likely to be harder to identify, particularly with regard to their source(s). Local pollution has been reduced dramatically in recent years, thanks to a concerted effort by the Landmark Trust/the Lundy Company, with advice from the Environment Agency. Gone is the practice of tipping incombustible rubbish over the cliff; and the run-off from island-generated sewage now receives treatment so that it is no longer a pollution hazard. However, there still remains the problem of pollution emanating from sources outside the area, including oil spills (oiled auks continued to be found along the strandline in the Landing Bay from time to time) and seaborne litter (particularly a problem after a spell of easterly winds). Unfortunately, little can be done on the island to prevent these from occurring.

WHAT PROTECTION CAN BE OFFERED?

Protecting a nature reserve on land is likely to involve a suite of measures designed to prevent unwanted predators or invaders from entering, and other measures, such as habitat creation or the removal of unwanted plant species, which could be classified as 'active management'. With a marine reserve, the term 'protection' needs to be viewed slightly differently. Clearly, it is impossible to 'fence in' an area of sea. Protection should be seen more in terms of the management of the area, what activities should be allowed (or not allowed) in particular parts of the reserve, and whether any form of 'active management' can be undertaken.

Byelaws

Byelaws are clearly the strongest deterrent for anyone intending to disturb the sea bed or destroy the marine life. Section 37 of the Wildlife and Countryside Act (1981) states that: 'without prejudice ... byelaws made under this section relating to a marine nature reserve may provide for prohibiting or restricting ... (i) the entry into, or movement within, the reserve of persons and vessels; (ii) the killing, taking, destruction, molestation or disturbance or animals or plants of any description in the reserve, or the doing of anything therein which will interfere with the sea bed or damage or disturb any object in the reserve; or (iii) the depositing of rubbish in the reserve.' This would appear to provide all the protection for the area that was needed. However, any prohibition or restriction on fishing activity would have to be made through the Devon Sea Fisheries Committee (DSFC). At that time, the DSFC were reluctant to introduce any new byelaws which would discriminate against any one type of fishing. This even included spearfishing. After careful negotiations however, the DSFC agreed to introduce a byelaw which would restrict the use of bottom gear (dredging/trawling) and tangle nets within the MNR.

More recently (2002), the DSFC introduced a byelaw preventing all fishing activity from taking place within the No-Take Zone off the island's east coast (see below).

Management of the area

The writing of a Management Plan was seen by the Nature Conservancy Council as being one of the first requirements in the move from voluntary to statutory status for the MNR. The first draft plan was compiled by Dr Keith Hiscock (Hiscock, 1983), then of the Field Studies Council's Oil Pollution Research Unit in Pembrokeshire and the main instigator of the scientific research which took place within the voluntary reserve during the 1970s and early 1980s. The Management Plan was re-written in 1994 (English Nature, 1994) and, for the first time, covered the island's Site of Special Scientific Interest (SSSI) as well as the MNR, as by this stage separating the two for management purposes was becoming increasingly difficult. The stated aim of the Plan (for the MNR) was 'to manage the MNR for the benefit of its wildlife, reconciling this with the sustainable use of its fisheries'. The five-year plan included a register of projects from which the annual work programme for the Warden was based. The Zoning Scheme was introduced soon afterwards in early 1995 (English Nature, 1995). This split the MNR into a number of 'usage zones' (see Figure 2 for the most recent version of the Zoning Scheme). The scheme, first pioneered in marine reserves abroad, helps to summarise the various byelaws covering the reserve in an easy-to-understand way.

Whilst the day-to-day running of the MNR/SAC is looked after by the Warden, the overall management is overseen by the Lundy Management Group. This body also has responsibility for nature conservation matters on the island itself. The Group consists of all Statutory Consultees (with respect to the island's management) and advisory organisations with specific interest in the management of the island and its environs (Lundy Management Group, draft Terms of Reference, 2006). Currently the Group consists of the following: Natural England, Devon Sea Fisheries Committee, Environment Agency, English Heritage, Landmark Trust, National Trust, RSPB, Lundy Field Society, MNR Advisory Group Chairman, and Defra RDS. The Group meets a minimum of twice a year following meetings of the MNR Advisory Group.

The Advisory Group acts as a forum for all those with an interest in the waters around Lundy. It restricts itself to just the marine reserve and does not discuss issues affecting the terrestrial part of the island. It provides an opportunity to discuss the day-to-day running of the MNR/SAC amongst those who actually use the island's waters, such as fishermen, divers, charter boat skippers and scientists. It also allows representatives of the Management Group to air new ideas and new policies. A report of the Advisory Group's discussions is published annually in the Lundy Field Society's Annual Report (for instance, see Irving, 2003)

The role of the Warden
Clearly the Warden plays an important part in the overall protection of the area, simply by being on site. With the provision of a patrol boat, visiting dive boats and other craft are able to see that a certain amount of 'policing' of the site is being done. Sadly though, with the other duties the Warden is expected to carry out, relatively little time is actually spent out on the water interacting with visitors to the MNR/SAC. Even with the help of an assistant during the summer months to help (amongst other duties) with the boat patrols, there will be long periods of time when there is no policing presence.

The Warden also acts as the Devon Sea Fisheries Committee's eyes and ears on the island, reporting any suspicious behaviour by fishing boats seen within the MNR/SAC, and within the No-Take Zone in particular. However, he/she has no powers of arrest or powers to confiscate gear and firm evidence of actual illegal fishing taking place would be required before any form of prosecution could be contemplated.

The No-Take Zone
The declaration of the statutory No-Take Zone (NTZ) in January 2003 has added another 'layer' of protection for habitats and species present off the island's east coast (Figure 2). The concept is well established in other countries (particularly New Zealand) but it is a relatively new idea for the U.K. It was first put to local fishermen in March 2001, but had been bandied about as an idea for at least three years prior to that. Essentially, all forms of fishing are prohibited from taking place within the NTZ.

The NTZ covers an area of approximately 8 km^2 of sea and is governed by a Devon Sea Fisheries byelaw which states, 'for marine environmental purposes, no person shall remove any sea fish from the area'. The hope is that the NTZ will have a number of long-term benefits including (i) increasing populations of fish and shellfish stocks within and outside the closed area; (ii) greater catches of fish for fishermen around the edges of the closed area; (iii) increasing the wealth of marine life, recreating the natural ecosystems and (iv) increasing benefits to local economies from tourism, diving and research. A comprehensive monitoring programme to study the impact of the NTZ began in 2004 and will continue until at least 2007. There are two parts to the monitoring programme: firstly the effects of the NTZ on commercial species (such as lobster, edible crab, spider crab, velvet crab and scallop); and secondly, the effect on long-lived sessile biota on rock habitats such as sea fans *Eunicella verrucosa*, erect sponges, ross or rose coral *Pentapora foliacea* and dead man's fingers *Alcyonium digitatum*.

During the consultation period prior to the designation of the NTZ, it was pointed out that as very little fishing effort was undertaken off the east side of the island, the benefits of introducing the NTZ may be quite difficult to determine.

However, Hoskin *et al.* (2006) have stated that, after two years of monitoring, initial indications are that:

(i) Lobsters appeared to have doubled in abundance (probably as a result of immigration of adult lobsters into the NTZ) and that their size was slightly larger within the NTZ.

(ii) None of the crab species had shown significant changes in terms of abundance or size.

(iii) The greater size of populations of several of the epifaunal species being monitored, when compared to populations outside the NTZ, probably reflects the situation before the NTZ was established.

(iv) Finally, the size of individual scallops was found to be significantly larger within the NTZ and they were also more abundant. However, these differences are likely to have originated prior to the designation of the NTZ.

Policing of the NTZ is the responsibility of the Devon Sea Fisheries Committee. There have been two known infringements by pot fishermen during 2005 and 2006, when pots have been set inside the NTZ boundary. In the first instance, the culprit was given a severe warning but escaped having a fine imposed. In the second instance, the culprit was anonymous and the 30 or so pots were confiscated.

Other protection measures

Ironically, before the declaration of the No-Take Zone, the two areas with the greatest protection for marine nature conservation purposes within the MNR/SAC were the exclusion zones around the two protected wrecks. Divers are prohibited from entering these two areas (extending to a radius of 100 m around the Gull Rock site, and for 50 m around the *Iona II*) unless they are in possession of a licence to dive them. They have therefore been the least disturbed sites around the island.

The Code of Conduct requests that boats do not anchor within 100 m of the Knoll Pins, and that dive boats are requested not to drop shot (weighted) lines there too, on account of the high concentration of delicate marine life. The concern that has been raised by the potential damage that anchors can do to the seabed has led to a flexible mooring being set in place in Gannets Bay close to the north-east corner of the island in 2005. The mooring can best be described as resembling a buoy fixed to an immovable concrete base by means of a giant rubber band. The benefit of this system is that it avoids the sunken part of a mooring chain being swept around on the seabed and damaging the biota.

Three species which occur within the MNR/SAC have greater protection than all others. These are the pink sea fan *Eunicella verrucosa* (see Box 2) and the basking shark *Cetorhinus maximus* (listed on Annex V of the Wildlife and Countryside Act, 1981); and the grey seal *Halichoerus grypus* (notified as an Annex II species present as a qualifying feature under the Habitats Directive, 1992).

Occasionally, hands-on active management can be undertaken to remove what could be termed a 'threat' to the area. This happened in the spring of 2005 with the removal of young 'japweed' *Sargassum muticum* plants from the Landing Bay area by volunteer divers from the Appledore Sub-Aqua Club. Japweed is a non-native species of brown seaweed originating from the N.W. Pacific which is capable of out-competing native seaweeds for space in the shallow sublittoral and in low shore rockpools. Whilst it is unlikely that it could be eradicated from the MNR/SAC entirely, it is important that the rockpools in the Devil's Kitchen (which are part of the SAC intertidal reefs monitoring programme) remain free of this invasive seaweed.

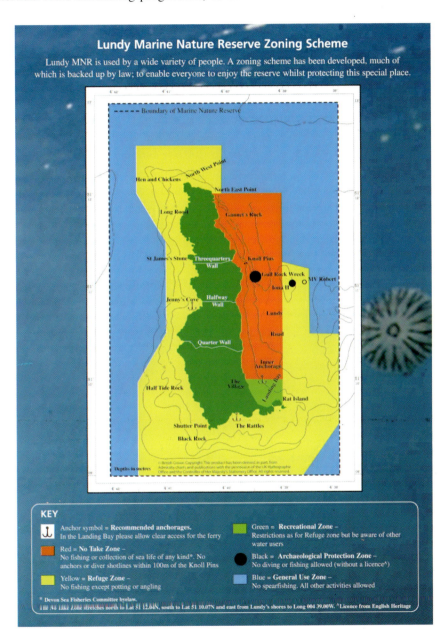

Figure 2: The Lundy Marine Nature Reserve Zoning Scheme (2003). The No-Take Zone is shown in red. The outer (hatched) line marks the boundary to both the MNR and the SAC.

MONITORING CHANGES TO THE MARINE LIFE

Appropriate management of the MNR/SAC depends, in part, on adequate knowledge about recruitment and longevity in species of nature conservation importance (Hiscock, 1994). Prior to 1984, little was known about the ecology and life history of many of these species, or whether they were particularly sensitive to changes or impacts. The monitoring programme, which ran from 1984 to 1991, was designed to address some of these unknowns. Results from this monitoring programme revealed that many of the most interesting species are very long-lived, but only recruit intermittently (Fowler & Pilley, 1992). Overall trends in their abundance are downwards (see Box 3), but what is unclear is whether this is merely a temporary downturn or whether it is part of a long-term pattern. On-going monitoring includes recording seawater temperatures throughout the year using automatic data loggers (Plate 1). A comprehensive programme monitoring the intertidal and subtidal reef 'features' around the island took place in 2003/4, forming a baseline study for assessing the overall condition of reefs within the SAC (Mercer *et al.*, 2006).

Clearly it is important to know if changes are taking place, whether these are for the better or the worse. Detecting such change can be extremely difficult, particularly as changes may be very subtle and may take several years to manifest themselves. Very little management intervention is possible however, should one wish to remedy a worsening situation.

Plate 1: Divers attaching an automatic temperature logger to the superstructure of the wreck of the M.V. *Robert*, off Lundy's east coast. *Photo: Paul Kay*

WHY PROMOTE THE AREA?

As part of its role as the Government's nature conservation advisor for England, Natural England (and its predecessors English Nature and the Nature Conservancy Council) has an obligation to identify and promote the best examples of marine habitats and species within the country. Putting on view something which might be better off being hidden presents somewhat of a dilemma. Some would argue that one of the best ways of protecting an area is to tell people why it is special, educate them in understanding which habitats or species are vulnerable, and hoping that by so doing they will 'take care' of the area and the wildlife within it. An example of this approach can be taken from the late 1960s. At this time (as mentioned earlier

Box 1: The Fall and Rise of the Red Band Fish at Lundy

The red band fish *Cepola rubescens* is a bottom-dwelling species found in areas of muddy gravel. This muddy-gravel seabed type usually occurs in depths of 70 m or more further offshore, but at Lundy this habitat type is present off the east side of the island in depths of 12-22 m (below chart datum). The eel-shaped fish, up to 70 cm long, spends much of its time hidden within a vertical burrow, emerging only to feed on passing plankton or to 'socialise' with others within harem-type groups.

In 1977, the population of these fish at Lundy was estimated to be about 14,000 individuals (Pullin & Atkinson, 1978). However, by 1983 not a single fish nor a single burrow could be found: the whole population appeared to have completely disappeared. Towed diver searches continued on an annual basis but it was not until 1987 that a small number of burrows (some with fish in them) were re-discovered (Irving, 1989). Since then the population has steadily grown, but numbers are still far less than they were in the late 1970s.

It is not known what may have caused this sudden decline in numbers. Atkinson *et al.* (1977) reported that there appeared to be a constant recruitment of young *Cepola* into the Lundy population, though there was also a high proportion of older fish present (9-12 yr old cohort), many of which may have died of natural causes within a short space of time. The recruitment of young fish may not have been able to continue had there been a high mortality of older fish. Alternatively, a mass mortality event may have occurred (the cause of which is unknown); or a disturbance of some sort may have caused the whole population to move away from the island; or there may be some other cause.

Plate 2: The anterior 12 cm of a red band fish *Cepola rubescens* appearing out of its vertical burrow. Note another burrow is present bottom right of the photograph. Photo taken in Halfway Wall Bay in 1987.

Photo: Robert Irving

Plate 3: Two thirds of the length of a male red band fish emerging from its burrow. Males have a distinctive iridescent blue colouration to their dorsal and ventral fins. Photo taken in Halfway Wall Bay in 1987.

Photo: Robert Irving

Box 2: The case of the sickly sea fans

The pink sea fan *Eunicella verrucosa* (Plate 4) is protected under Appendix V of the Wildlife and Countryside Act 1981 against killing, injuring, taking possession and sale. Since 1999, it also has had its own Biodiversity Species Action Plan, recognising its status as being rare and vulnerable. It is found throughout south-west Britain, from Portland (Dorset) to north Pembrokeshire, as well as in southern Ireland.

Monitoring studies of the sea fans at Lundy, which started in 1984, showed a marked deterioration in the overall condition of Lundy's sea fans from 2000 to 2002 in particular (Irving & Northen 2004). Indeed, when compared to the condition of sea fans from other sites, the population at Lundy was shown to be in a particularly poor state of health (see Fig. 3). The cause of this decline in condition was clearly not due to physical disturbance as individual fans remained attached and partially living. Instead it was thought that some change in water quality was affecting the fans. Studies at Plymouth University and at the Marine Biological Association have since shown that a bacterium *Vibrio splendidus* was attacking the soft tissue of the fans (pers. comm., J. Hall-Spencer) (Plate 5). It is believed that the infection has now passed through the population, as individual sea fans appear to be recovering (pers. comm., C. Wood).

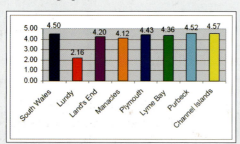

Figure 3: Chart showing the average 'condition' of seafans at eight sites in southern Britain during 2001/02. 'Pristine' condition scores 5, 'sickly' condition scores 1. (After Wood, 2003).

Plate 4. A sea fan *Eunicella verrucosa* in 'pristine' condition at Lundy in 1997. The size of this fan is approximately 25 cm x 25 cm.

Photo: Paul Kay

Plate 5. Close-up view of part of a 'sickly' sea fan at Lundy in 2000, overgrown by barnacles and bryozoan turf.

Photo: Robert Irving

Box 3: The decline in the population of sunset cup corals

In the U.K., the sunset cup coral *Leptopsammia pruvoti* is a species of particular marine natural heritage importance: it is nationally rare and has its own Biodiversity Action Plan (BAP). As a Mediterranean-Atlantic species, *L. pruvoti* is at the northern extreme of its range at Lundy. It is only found at a handful of other sites in south-west Britain: the Isles of Scilly, off Plymouth Sound, in Lyme Bay and at Portland Bill. Within these populations there appears to be very little new recruitment in evidence and, consequently, the number of individuals is declining. This is of particular concern to conservationists. The population of *L. pruvoti* re-photographed at the Knoll Pins on an annual basis between 1983 and 1990 was found to have lost 8% of its individual corals (Fowler & Pilley, 1992), and between 1984 and 1996 part of this same population had declined by 22% (Hiscock, 2003).

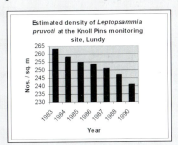

Figure 4: Diagrammatic representation of the decline in numbers of *Leptopsammia pruvoti* from 1983 to 1990 from a population at the Knoll Pins (after Fowler & Pilley, 1992).

The cause of this decline is not known for certain, but studies undertaken at Lundy indicate that the presence of the horseshoe worm *Phoronis hippocrepia* around the base of several individual cup corals could be important (Irving, 2004). These worms bore into the calcium carbonate skeleton of the cup coral, thereby weakening its attachment to the underlying rock and possibly causing it to fall off the rock. Several skeletons of dead cup corals, collected at the foot of cliffs below *Leptopsammia* populations have been found to have worm-borings in their bases.

Plate 6: A cluster of sunset cup corals *Leptopsammia pruvoti*, photographed at the Knoll Pins.

Photo: Paul Kay

Plate 7: A *Leptopsammia pruvoti* individual with the lophophores of several *Phoronis hippocrepia* horse-shoe worms emerging from its base.

Photo: Robert Irving

in this paper), sea fans used to be collected as souvenirs at Lundy. Those divers collecting them probably had no idea that the sea fans' growth rate of just 10 mm/year meant that an average sized specimen measuring 50 cm in height was at least 50 years old. By explaining this fact to divers, it became much easier to persuade them not to collect sea fans, and the practice was soon ended.

However, by promoting the area one would expect more people to visit than might otherwise have done so. Thus there is an increased risk of damage to the very thing one is trying to protect, simply by an increase in 'visitor pressure' (see Box 4). A profit-making business may well wish to advertise its wares in order to get more people to buy whatever it might be trying to sell. But nature conservation organisations do not work like that: there is a genuine desire to want to inform and educate people for the ultimate benefit of wildlife. This is particularly the case with underwater habitats and species which, for the majority of people, remain as total unknowns and outside their own personal experiences. Even divers are often unaware of the biology of, and conservation interest in, many of the species they see under water.

Box 4: The Gull Rock wreck site

In 1968, some stone cannonballs and iron cannon were found in the vicinity of Gull Rock by John Shaw, one of Lundy's diving pioneers. The artefacts were well camouflaged and the site was not found again until 1983. Since then a series of surveys have shed light on what may be Lundy's most important underwater site (Robertson & Heath, 1997). It is thought that the artefacts date from the 15th or 16th centuries, but it is not certain how they came to be there - was there a shipwreck (no evidence of one has been found) or were they jettisoned from a ship? Whatever their origin, in 1989 the Archaeological Diving Unit decided that the remains were of national importance and the site was designated in 1990 under the Protection of Wrecks Act, 1973.

When the site was first discovered, it was realised that the artefacts were likely to generate considerable interest amongst 'wreck hunters' and so the location of the site was kept a closely guarded secret. Ironically, with the notification of the Protected Wreck status and the 100m diameter exclusion zone, the location of the artefacts became public knowledge and within a matter of months, several of the artefacts had mysteriously disappeared.

One wonders whether, in this case, informing all and sundry about the importance of the site prior to its announcement as a protected wreck would have avoided the looting of the site (for that is probably what has happened). An alternative solution, though one that would be far more expensive, would be to remove the artefacts from the seabed and conserve them by the appropriate means at a safe location. This solution would involve a long-term commitment of funding, time and space, and satisfying all three requirements would seem to be extremely unlikely.

There is also the 'kudos' factor - the more people that visit an area, the more it can be said to be popular and appealing. This can help with attracting more people to visit the island and with grant applications (grant-awarding bodies often require some indication of visitor numbers to a site).

HOW IS THE AREA BEING PROMOTED?

One of the most beneficial things to come about from the establishment of the voluntary marine nature reserve at Lundy was the six-month appointment of the country's first marine Warden in 1978. This was viewed as a pilot project aimed mainly at assessing the work of a marine warden. As part of his brief he was asked to:

1. ensure the Code of Conduct was abided by;
2. provide guidance and information to visiting divers;
3. assist field workers in carrying out their studies;
4. assist in organising and running field courses in sublittoral ecology; and
5. prepare illustrated guides for the reserve.

As part of point (5) above, an illustrated guide to an underwater nature trail at the Knoll Pins was prepared. Whilst being regarded as an excellent idea, sadly the popularity of this experiment proved to be its downfall, as it encouraged divers to follow a set route around the site leading to the very marine life they had come to see being inadvertently damaged.

Despite the success of this pilot project, there were insufficient funds to continue it and a further eight years had to pass before the next marine warden was appointed after the designation of the statutory MNR. Since then there have been a further five incumbents, each bringing something new to the role. The post of Warden has been fundamental to the success of the MNR/SAC. Not only have they acted as a point of contact for the MNR/SAC on the island, they have also been instrumental in the educational promotion of the area for divers and non-divers alike. Illustrated talks are given on a weekly basis to those staying on the island; there are rockpool rambles in the Devil's Kitchen; and the snorkel trail between the jetty and Rat Island has proved a great success.

Table 3: A selection of initiatives used to promote Lundy's MNR/SAC.

PROMOTIONAL INITIATIVES	COMMENT
Information panels	All-weather promotional panels at Bideford and Ilfracombe - the main crossing points to Lundy. Information panels on board the M.S. *Oldenburg* and as part of the display area at the back of the Church on Lundy. Information panels in the Beach Building on Lundy.
Literature	ID book: The Scuba Diver's Guide to the Lundy Marine Nature Reserve. Numerous leaflets. Schools educational packs to fit in with various curricula.

Video/Media	Numerous TV and radio clips and some full-length programmes about Lundy's marine life.
	1996: first video about the MNR using donated footage.
	2002: EN-commissioned video about the MNR 'Lundy - an island to treasure' shown on board the M.S. Oldenburg during crossings to the island.
Web cam and virtual web tour	Underwater images from the Landing Bay broadcast on EN's website.
	Interactive web tour of the Lundy MNR on EN's website.

CONCLUSION

A great deal has been written over the past 30 years extolling the merits of Lundy's marine life and underwater scenery. The increase in interest in the site during this time has been accompanied by a considerable amount of marine biological research taking place, leading to a corresponding increase in our knowledge of the island's seabed habitats and marine life. The place has not been swamped by divers as was once feared would happen (its isolation has seen to that), and yet the amount of educational material about the marine life has increased considerably.

One can try to imagine what the waters around the island would be like today had no marine nature reserve been established. It is likely that the level of potting would have found its own equilibrium (probably similar to the level that can be sustained at present), though this may have come about after cycles of boom and bust years. It is highly likely that the soft sediments off the east coast would have been dredged for scallops and possibly other species, destroying the scientific interest of that particular area. The number of visiting divers would probably have been similar to the number which visit at present, though there would have been no restrictions on their activities.

On balance, one could conclude that the various designations have certainly been beneficial to the island's marine habitats and wildlife. Given the resources available to the statutory bodies responsible for the site's management, there will always be gaps in the protection that these designations should offer. However, it would be welcome if more resources could be put into patrolling the island's near-shore waters, and more emphasis be placed on a research programme assessing the declining fortunes of certain of the species of high nature conservation interest. It is appreciated, though, that little can be done to protect against external influences which may be detrimental to important species or communities. The management requirements of a marine reserve are clearly different to those of a nature reserve on land, with far less emphasis on human interference and more on long-term monitoring to distinguish natural trends from anthropogenic-generated anomalies. Finally, the long-term success or failure of the No-Take Zone will have a considerable bearing on the way fisheries interests are managed within marine reserves in future.

Please note that the views expressed in this paper are those of the author alone.

REFERENCES

Atkinson, R.J.A., Pullin, R.S.V. and Dipper, F.A. 1977. Studies of the red band fish *Cepola rubescens* L. *Journal of Zoology*, 182, 369-384.

Cole, P.B.F. 1986. The Lundy marine nature reserve. *Annual Report of the Lundy Field Society 1985*, 36, 44.

Duck, C.D. 1996. Chapter 5.14 Seals. *In: Coasts and seas of the United Kingdom. Region 11 The Western Approaches: Falmouth Bay to Kenfig*, ed. By J.H. Barne, C.F. Robson, S.S. Kaznowska, J.P. Doody, N.C. Davidson & A.L. Buck, 146-148. Peterborough, Joint Nature Conservation Committee. (Coastal Directory Series).

English Nature. 1993. *Conserving England's marine heritage - a strategy*. Peterborough: English Nature.

English Nature. 1994. *Managing Lundy's Wildlife - A Management Plan for the Marine Nature Reserve and the SSSI*. Unpublished report, English Nature Maritime Team, Peterborough.

English Nature. 1995. *Lundy Marine Nature Reserve Zoning Scheme*. Peterborough: English Nature.

Fowler, S.L. & Pilley, G.M. 1992. *Report on the Lundy and Isles of Scilly marine monitoring programmes 1984 to 1991*. Report to English Nature from The Nature Conservation Bureau Ltd. Peterborough, English Nature.

Hiscock, K. 1972. The proposal to establish a marine nature reserve around Lundy - progress. *Annual Report of the Lundy Field Society 1971*, 22, 31-34.

Hiscock, K., Grainger I.G., Lamerton, J.F., Dawkins, H.D. and Langham, A.F. 1973. Lundy Marine Nature Reserve. A policy for the management of the shore and seabed around Lundy. *Annual Report of the Lundy Field Society 1972*, 23, 39-45.

Hiscock, K. 1982. *Lundy Marine Nature Reserve management plan. Supplement 1. Information file on the littoral ecology of Lundy*. Unpublished report to the Nature Conservancy Council, Huntingdon by the Field Studies Council, Oil Pollution Research Unit, Pembroke.

Hiscock, K. 1983. *Lundy Marine Nature Reserve Draft Management Plan*. Huntingdon: Nature Conservancy Council, iv, 87 pp.

Hiscock, K. 1994. Marine communities at Lundy - origins, longevity and change. *Biological Journal of the Linnean Society*, 51, 183-188.

Hiscock, K. 1997. Marine Biological Research at Lundy. In: R.A. Irving, A.J. Schofield & C.J. Webster (eds.), *Island Studies - fifty years of the Lundy Field Society*, 165-183. Bideford: Lundy Field Society.

Hiscock K. 2003. Changes in the marine life of Lundy. *Annual Report of the Lundy Field Society 2002*, 52, 84-93.

Hoskin, M., Coleman, R., Hiscock, K. and von Carlshausen, J. 2006. *Monitoring the Lundy NTZ: 2003 to 2005*. Unpublished draft report to English Nature, Defra and the World Wide Fund for Nature.

Irving, R.A. 1989. Searches for the red band fish *Cepola rubescens* L. at Lundy, 1984-1988. *Annual Report of the Lundy Field Society 1988*, 40, 53-59.

Irving, R.A. 2003. Report of the Lundy Marine Nature Reserve Advisory Group 2002. *Annual Report of the Lundy Field Society 2002*, 52, 99-102.

Irving, R.A. 2004. *Leptopsammia pruvoti* at Lundy - teetering on the brink? *Porcupine Marine Natural History Society Newsletter*, 15: 29-34.

Irving, R.A. 2005. *Lundy cSAC and MNR Literature Review*. Unpublished report to English Nature (Exeter) by Sea-Scope, Marine Environmental Consultants, Bampton, Devon.

Irving, R.A. and Gilliland, P. 1997. Lundy's Marine Nature Reserve - a short history. In: R.A. Irving, A.J. Schofield & C.J. Webster (eds.), *Island Studies - fifty years of the Lundy Field Society*, 185-203. Bideford: Lundy Field Society.

Irving, R.A. and Northen, K.O. (eds.) 2004. *Report of the MCS Working Parties to Lundy, 1997-2001*. Unpublished report to English Nature (Devon Team), Exeter.

Kay, P. 2002. *The Scuba Diver's Guide to Lundy Marine Nature Reserve*. Peterborough: English Nature. 60 pp.

Mercer, T.S., Howson, C.M. and Bunker, F. St P.D. 2006. *Lundy European Marine Site Sublittoral Monitoring 2003/4*. Unpublished report to English Nature, Peterborough by Aquatic Survey and Monitoring Ltd. EN contract FST20/46/16. Nature Conservancy Council. 1986. *Lundy Marine Nature Reserve*. [Map of statutory MNR boundary].

Nature Conservancy Council. 1987. *Lundy Marine Nature Reserve: Code of Conduct and Byelaws*. Leaflet produced by the Nature Conservancy Council, Peterborough.

Pullin, R.S.V. and Atkinson, R.J.A. 1978. *The status of the red band fish* Cepola rubescens *L. at Lundy*. Unpublished report to the Nature Conservancy Council, Huntingdon.

Robertson, P. and Heath, J. 1997. Marine Archaeology and Lundy. In: R.A. Irving, A.J. Schofield & C.J. Webster (eds.), *Island Studies - fifty years of the Lundy Field Society*, 77-86. Bideford: Lundy Field Society.

Wood, C. 2003. *Pink sea fan survey 2001/2*. Unpublished report for the Marine Conservation Society, Ross-on-Wye, 23 May 2007.

LUNDY'S LENTIC WATERS: THEIR BIOLOGY AND ECOLOGY

by

JENNIFER GEORGE

Sabella, Gays Lane, Holyport, Maidenhead, Berks, SL6 2HL
e-mail: georgej@wmin.ac.uk

ABSTRACT

Research on the Lundy freshwater ecosystems in the late 1970s, 1980s and early 1990s showed that the major standing water bodies (lentic waters) supported different populations of organisms, particularly in the planktonic and macroinvertebrate groups. Recent research in the autumn of 2003, spring 2005 and winter 2006 not only demonstrated that these differences still are present, but also gained information on seasonal changes occurring in these waters. Four water bodies, Pondsbury, the Rocket Pole Pond, Quarry Pool and the larger pond at Quarter Wall have been studied at all seasons. Differences can be related to the position of the pond on the island and hence degree of exposure, the amount of plant cover and the input of decaying material and nutrients. At various times during the last 27 years smaller, often temporary, bodies of water have been surveyed e.g. pools in North Quarry, smaller pond at Quarter Wall, pond in Barton Field. Brief descriptions of these small ecosystems are given here.

Keywords: *Lundy, ponds, aquatic plants, plankton, macroinvertebrates*

INTRODUCTION

Freshwater ecosystems can broadly be separated into two categories, the moving or lotic waters and the still or lentic waters. Both types occur on Lundy and the positions of these water bodies were documented by Langham (1969). The streams (lotic) systems had received practically no attention until Long carried out a comprehensive survey in the summer of 1993. He showed that the stream fauna was impoverished compared to similar streams on the mainland (Long, 1994). The lentic waters can further be divided into the permanent or semi-permanent type such as Pondsbury, the Rocket Pole Pond, Quarry Pool and the larger pond at Quarter Wall, and the smaller seasonal often temporary pools such as the smaller pond at Quarter Wall, the pools in the quarries and in the depressions to the east of the Rocket Pole.

Until 1979 there had been no detailed investigation of the lentic waters, although there had been studies of individual freshwater organisms e.g. Hemiptera (Morgan, 1948), diatoms (Fraser-Bastow, 1950), Crustacea and Rotifera (Galliford, 1954), isopod *Asellus* (Williams, 1962), fish (Baillie and Rogers, 1977). Some freshwater organisms have been listed in surveys of terrestrial groups such as Coleoptera

(Brendell, 1976), Diptera (Lane, 1978), Hemiptera (Alexander, 1992) and in the lists of invertebrates, compiled by Parsons, in the Lundy Field Society's Annual Reports since 1982.

In the summers of 1979 and 1986 investigations of the biology and ecology of the four main bodies of water, Pondsbury, the Rocket Pole Pond, Quarry Pool and the larger Quarter Wall pond were carried out (George and Stone, 1980, 1981; George and Sheridan, 1987). Further detailed information of Pondsbury was obtained by Clabburn in his comprehensive summer survey in 1993 (Clabburn, 1994). All of these data which relate to summer conditions were reviewed by George (1997) who highlighted the need for comparative seasonal information. In the autumn (mid-October) of 2003 further investigations were carried out on the four main lentic waters (George, McHardy and George, 2004; George, McHardy and Hedger, 2004) and in the spring 2005 (April) and winter 2006 (January) which has given a seasonal perspective to the ecology of these waters. Also some of the smaller temporary pools were examined in these surveys.

This review will consider the ecology of the four permanent water bodies, Pondsbury, the Rocket Pole Pond, Quarry Pool and the larger Quarter Wall pond with a view to assessing the comparative stability of these waters over the past 27 years together with data on seasonal changes. Brief discussion on some of the temporary waters will be included.

METHODS
Physical and chemical measurements
The following factors were measured at each pond: air and water temperatures, pH (pH meter), oxygen content of surface and bottom water (oxygen meter). Mapping of the ponds was undertaken and depth measurements were also plotted.

Flora
The species of plants within and at the edges of the ponds were listed and the distribution and location of the main species were plotted on to outline maps of the ponds. In all of the seasonal surveys a subjective estimate of relative abundance of each of the species was made on a scale of 1-5 as follows:

Score	Relative Abundance	
1	Rare	Less than 1% of total number of plants present
2	Occasional	1-5% of total number of plants present
3	Frequent	6-10% of total number of plants present
4	Common	11-50% of total number of plants present
5	Abundant	More than 50% of total number of plants present

At Pondsbury plants were surveyed at 50 sites around the perimeter and recorded as present or absent and abundance noted. Relative abundance was recorded for each species by taking an average for relative abundance at all sites.

Plankton
Plankton was collected with a FBA phytoplankton net (aperture 0.075mm) and in the Rocket Pole Pond, Quarry Pool and the Quarter Wall pond two hauls were taken across each pond. At Pondsbury two hauls were taken from the mid-northern side to the mid-

west side across the deeper part of the water body. Samples were fixed in 4% formaldehyde and transferred to ethanol for microscopic examination in the laboratory.

An estimate of relative abundance of each taxon was made on a scale of 1 to 5 as follows:

Score
1	*One or two only of the taxon*
2	*3-25 of the taxon*
3	*26-100 of the taxon*
4	*101-500 of the taxon*
5	*Over 500 of the taxon*

Macroinvertebrates

Macroinvertebrates were collected from the plant beds and open water using a standard FBA net (aperture 0.96mm) by sweeping for five one-minute periods at each pond. No quantitative sampling was undertaken in the sediments as preliminary sampling showed that no organisms occurred that were not represented in the sweep samples.

An estimate of relative abundance of each taxon on a scale of 1 to 5 was made as follows:

Score
1	*Less than 5 individuals*
2	*5-49 individuals*
3	*50-199 individuals*
4	*200-499 individuals*
5	*Over 500 individuals*

ORIGINS AND CHARACTERISTICS OF THE WATER BODIES

Pondsbury (OS Grid reference SS 13463 45508) which is the largest body of freshwater on the island is surrounded by *Sphagnum* bog, heathland and rough grazing pasture. It is probably of natural origin although the construction of an impounding bank on its west side has increased its size and depth. It receives surface run-off from the surrounding land and it has an outlet situated midway along the raised bank which forms the Punchbowl stream that flows into the sea at Jenny's Cove. Detailed mapping of the water body shows that it regularly changes shape due to varying water levels, macrophyte encroachment, silt deposition and human activity such as dredging and damming. Its area during the six survey periods from 1979-2006 varied from 3300m^2 to 4000m^2, although records show that in the past, e.g. summer of 1976, it dried up altogether.

The other three water bodies which have been formed from excavations in the rock and are smaller in size, maintain their shape, although their water levels fluctuate according to weather conditions.

The Rocket Pole Pond (OS Grid Reference SS 13481 43681) is a steep-sided water body cut into granite, 25m x 11.5m in size with its western side stepped and much shallower. There is no through drainage. It is fully exposed to the westerly winds that commonly blow across the island.

Quarry Pool (OS Grid Reference SS 13756 95037) is a very sheltered body of water, 22m in length and 11m wide, overshadowed by steep rocky walls and some trees e.g. willow. It is fed by a small stream falling over granite boulders and has an outlet on its eastern side.

The larger **Quarter Wall pond** (OS Grid Reference SS 13630 44965) is an open body of water, 19m x 12m in size, with fairly steep rocky banks. It is situated at a fairly high level on the island's eastern side, and probably receives little surface drainage.

RESULTS: PONDSBURY

The physical and chemical characteristics, flora and fauna of Pondsbury, together with seasonal changes will be given first of all as this water body is much larger with a greater biodiversity than the other three similarly-sized ponds.

Physical and chemical characteristics

Maximum depths occur in the northern part of Pondsbury where depths of over one metre were recorded in all the seasonal surveys (Table 1). Dredging of this area took place in 1993 and 1995. To the east and south the water becomes progressively shallower. Water temperatures relate to the ambient air temperatures at the time of sampling. Temperature recording over a 24h period showed that maximum water temperatures occur in the late afternoon/early evening with minimum temperatures in the early morning (Clabburn, 1994). Dissolved oxygen values show that at all seasons the surface waters are well-oxygenated. 24h recording showed that considerable fluctuations in oxygen take place just above the sediments (Clabburn, 1994). Pondsbury is acidic with the pH ranging from 4.8 in summer to 6.4 in winter.

Table 1: Physical and chemical measurements recorded in the north-east section of Pondsbury

	SPRING 2005	SUMMER 1986	1993	AUTUMN 2003	WINTER 2006
Max. depth m.	1.65	1.0	1.2	1.3	1.8
Air temp. °C	14	17	-	11	5.5

Water temp.
  Surface °C
  Bottom °C
	12
	12
		15.5
		15
			14.4*
			14.5*
				12.5
				12
					3.5
					3.5

pH	6.0	4.8	4.82	5.0	6.4

Oxygen % satn.
  Surface
  Bottom
	95
	85
		86
		77
			76.9*
			54.4*
				97
				80
					-
					-

Key: *Values represent the mean of diel fluctuations recorded (Clabburn, 1994).

Flora

Fifteen plant species have been found in and around the margins of Pondsbury, although their relative abundance varies with the time of year (Table 2). The moss, *Sphagnum cuspidatum*, dominates the entire area and the water body is surrounded by large stands of the rush, *Juncus effusus*, particularly on its western side. Another rush, *Eleocharis palustris*, which dies down during the winter months is fairly widespread, and it forms stands further out into the open water than *Juncus effusus*. Common submerged species present in all seasons were *Potamogeton polygonifolius*, the bog pondweed, *Hydrocotyle vulgaris*, water pennywort, *Callitriche stagnalis*, the mud water starwort, and *Myosotis scorpioides*, the creeping water forget-me-not. *Hypericum elodes*, marsh St John's wort. which dominated the entire southern area of Pondsbury in 1979 was far less abundant in the 2003, 2005 and 2006 surveys probably due to the dredging that occurred in the autumn of 1993 and 1995.

Table 2: Species present and relative abundance of the flora in Pondsbury

SPECIES	SPRING 2005	SUMMER 1979	SUMMER 1986	AUTUMN 2003	WINTER 2006
Sphagnum cuspidatum Ehrb.	5	5	5	5	5
Hypericum elodes L.	3	5	5	2	-
Hydrocotyle vulgaris L.	3	5	5	3	2
Ranunculus flammula L.	-	2	2	1	-
Ranunculus omiophyllus Ten.	-	1	1	-	-
Callitriche stagnalis Scop.	4	2	2	4	3
Potamogeton polygonifolius Pourret	4	1	3	4	4
Juncus effusus L.	5	4	5	5	5
Juncus articulatus L.	2	3	2	3	2
Juncus conglomeratus L.	1	-	-	1	-
Eleocharis palustris L.	4	2	3	4	4
Myosotis scorpioides L.	4	5	4	3	4
Lythrum portula L.	-	2	3	-	-
Elatine hexandra (Lapierre)	-	1	1	-	-
Galium palustre L.	-	-	1	-	-
TOTAL SPECIES PRESENT 15	10	13	14	11	8

Key: 5: >50%; 4: 11-50%; 3: 6-10%; 2: 1-5%; 1: <1% of total number of plants present.

Plankton

Pondsbury showed good diversity in its plankton population with 12 species of phytoplankton and 17 species of zooplankton being recorded from the seasonal surveys (Table 3). Figure 1 shows the seasonal differences in the main planktonic groups. The Cladocera (water fleas) were present throughout the year, with *Daphnia*

Table 3: Organisms in the Pondsbury Plankton

PHYTOPLANKTON	ZOOPLANKTON
EUGLENOPHYTA *Euglena viridis* Ehrb.	ANNELIDA - OLIGOCHAETA *Nais* sp.
CHLOROPHYTA (Green algae) *Volvox* sp. *Desmodesmus(Scenedesmus) magnus* Chodat *Ankistrodemus* sp. *Coelastrum* sp. *Closterium* sp. *Cosmarium* sp. Filamentous green alga BACILLARIOPHYTA (Diatoms) *Melosira* sp. *Tabellaria* sp. *Pinnularia* sp. *Diatoma* sp.	CRUSTACEA - CLADOCERA *Daphnia obtusa* Kurz *Bosmina longirostris* (Muller) *Chydorus sphaericus* (Muller) *Alonella nana* (Baird) CRUSTACEA - COPEPODA *Cyclops* sp. Immature cyclopids Cyclopid nauplii larvae Harpacticoids CRUSTACEA- OSTRACOD Cypridid ostracod ROTIFERA *Keratella vulga* (Ehrb.) *Keratella quadrata* (Muller) *Keratella serrulata* (Ehrb.) *Euchlanis dilatata* (Ehrb.) *Northolca acuminata* (Ehrb.) *Cephalodella* sp. INSECTA - EPHEMEROPTERA *Chloeon dipterum* (L.) l. INSECTA - DIPTERA *Chaoborus crystallinus* (Deg) l. Chironominae l.

obtusa being recorded in all four seasons. In the winter samples of 2006 this species constituted 75% of the total number of organisms present. The Copepoda (copepods), particularly *Cyclops* sp. were also well represented in the plankton, with all life cycle stages, larvae, immature cyclopids, adults, being present throughout the year. The Rotifera (rotifers) occurred at all seasons but were represented by different species e.g. *Keratella quadrata* was abundant in the spring, *Keratella serrulata* in the summer months and *Keratella vulga* in the autumn. Two species appeared in large numbers at particular seasons; the green alga, *Euglena viridis* produced a large bloom in the summer and the larva of the phantom midge, *Chaoborus crystallinus* appeared in large numbers in the autumn.

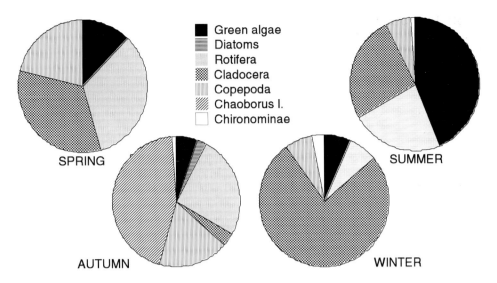

Figure 1: Seasonal differences in Pondsbury plankton

Macroinvertebrates

The species and abundance of macroinvertebrates in the sweep samples are given in Table 4 where rare species (scale 1) found only on one occasion are not shown. The greatest diversity occurred in the summer months and Clabburn (1994) in his comprehensive survey of Pondsbury (10 sites with 2-minute sweep samples at each) recorded 35 macroinvertebrate species with a further nine identified to generic level. The most abundant macroinvertebrate in Pondsbury was the isopod crustacean, *Asellus meridianus* which occurred throughout the year, although being far less abundant during the winter months. In the spring samples ovigerous females and many small individuals were found. Other species which were found at all seasons were the flatworm, *Polycelis nigra*, the oligochaete, *Lumbriculus variegatus*, the lesser water boatman, *Corixa punctata* (two comatose individuals were found in the plant beds at water temperatures of 3.5°C in January 2006) and various chironomid larvae species. Very few species were found in the winter samples due to the extremely cold water conditions and it is likely that most were lying dormant amongst the roots of the plants or in the sediments. *Argyroneta aquatica*, the water spider. which was first recorded by Galliford in 1953, and seen again in 1979, 1986 and 1993 was again recorded in April 2005 where it was found amongst the *Sphagnum* on the south side of Pondsbury.

RESULTS: ROCKET POLE POND, QUARRY POOL, QUARTER WALL POND
Physical and chemical characteristics

Table 5 shows the physical and chemical characteristics of the three ponds together with seasonal differences. Depth recordings at the various seasons showed that the Rocket Pole Pond is the deepest water body, reaching a maximum depth of 2.2m in the summer survey of 1986. Quarry Pool appears to maintain a fairly constant body of water throughout the year with maximum depths of 1.4-1.5m being recorded during the last three years. Quarter Wall pond is a shallower body of water with a maximum depth

Table 4: Abundance of macroinvertebrate species in Pondsbury

SPECIES	SPRING	SUMMER	AUTUMN	WINTER
Platyhelminthes:				
Polycelis nigra (Muller)	3	4	2	3
Oligochaeta				
Lumbriculus variegatus (Muller)	2	3	2	2
Hirudinea:				
Helobdella stagnalis (L.)	1	1	-	-
Arachnida:				
Argyroneta aquatica L.	1	2	-	-
Hydacarina	-	2	-	-
Crustacea:				
Asellus meridianus Racovitza	3	5	3	2
Insecta: Ephemeroptera:				
Cloeon dipterum (L.) l.	1	1	2	-
Insecta: Odonata:				
Ischnura elegans (van de Linden) l.	-	2	2	-
Sympetrum striolatum (Charpentier) l.		2	1	-
Insecta: Hemiptera:				
Notonecta marmorea viridis Delcourt	2	2	2	-
Corixa punctata (Illiger)	2	2	2	1
Callicorixa praeusta (Fieber)	-	2	-	-
Cymatia bondorffi (Sahlberg)	1	2	-	-
Sigara dorsalis (Leach)	-	2	-	-
Immature corixids/cymatids	2	3	-	-
Insecta: Coleoptera:				
Hydroporus pubescens (Gyllenhal) a.	-	1	1	-
Hygrotus inaequalis (Fab.) a.	2	2	1	-
Laccophilus minutus (L.) a.	1	2	1	-
Agabus bipustulatus L. a.	1	2	-	-
Insecta:Diptera:				
Chironominae l.	2	4	2	2
Mollusca:				
Pisidium personatum Malm	-	3	-	-
TOTAL SPECIES FOUND	13	*43	12	5

Key: Abundance scale: 5: >500; 4: 200-499; 3: 50-199; 2: 5-49; 1: <5 individuals. L=larva, a=adult. *Species (Abundance 1) found on only one occasion are not recorded in the table.

Table 5: Physical and chemical measurements recorded at the three ponds

ROCKET POLE POND

	SPRING 2005	SUMMER 1979	SUMMER 1986	AUTUMN 2003	WINTER 2006
Max. depth m.	1.8	1.9	2.2	1.65	1.9
Air temp.°C	13	18	18	12	5.5
Water temp °C Surface Bottom	12.5 12	15 15	17 12	13.5 13	5 5
pH	6	5	5.4	5.5	6.18
Oxygen % satn. Surface Bottom	90 86	82 78	77 65	120 91	- -

QUARRY POOL

	SPRING 2005	SUMMER 1979	SUMMER 1986	AUTUMN 2003	WINTER 2006
Max. depth m.	1.4	1.64	1.7	1.5	1.48
Air temp. °C	14	19	18	12	5
Water temp °C Surface Bottom	12 12	17 15	17 16	14 13	6 6
pH	6	5	5.8	5.5	6.8
Oxygen % satn. Surface Bottom	92 89	78 49	81 67	113 78	- -

QUARTER WALL POND

	SPRING 2005	SUMMER 1979	SUMMER 1986	AUTUMN 2003	WINTER 2006
Max. depth, m.	0.75	0.4	0.8	0.4	0.75
Air temp.°C	14	20.5	18	14.5	5
Water temp.°C Surface Bottom	12.5 12	18.5 17	18 17	14 13.5	6 6
pH	6	5	5.9	5.5	5.8
Oxygen % satn. Surface Bottom	96 94	101 98	104 98	102 98	- -

Table 6: Species present and relative abundance of the flora of the three ponds

ROCKET POLE POND

SPECIES	SPRING 2005	SUMMER 1979	1986	AUTUMN 2003	WINTER 2006
Hydrocotyle vulgaris L.	2	4	2	4	-
Juncus effusus L.	3	2	2	5	5
Juncus conglomeratus L.	1	-	-	2	1
Eleocharis palustris L.	2	4	4	4	2
Myosotis scorpioides L	1	-	-	1	1
TOTAL SPECIES PRESENT 5	5	3	3	5	4

QUARRY POOL

SPECIES	SPRING 2005	SUMMER 1979	1986	AUTUMN 2003	WINTER 2006
Fontinalis sp.	-	2	-	-	-
Hydrocotyle vulgaris L	2	-	-	-	-
Ranunculus flammula L	-	3	2	2	-
Callitriche stagnalis Scop.	-	1	1	-	-
Potamogeton polygonifolius Pourret.	-	5	3	-	-
Juncus effusus L.	3	2	2	3	3
Eleocharis palustris L.	2	2	2	3	2
Myosotis scorpioides L.	1	-	-	1	2
Lythrum portula L.	-	1	-	-	-
TOTAL SPECIES PRESENT 9	4	7	5	4	3

QUARTER WALL POND

SPECIES	SPRING 2005	SUMMER 1979	1986	AUTUMN 2003	WINTER 2006
Hydrocotyle vulgaris L.	1	5	3	1	-
Callitriche stagnalis Scop.	-	-	2	-	-
Potamogeton polygonifolius Pourret	-	-	2	-	-
Juncus effusus L.	4	2	3	3	3
Eleocharis palustris L.	3	5	3	4	3
Myosotis scorpioides L.	2	1	2	2	2
Lythrum portula L.	2	4	2	1	1
TOTAL SPECIES PRESENT 7	5	5	7	5	4

Key: 5: >50%; 4: 11-50%; 3: 6-10%; 2: 1-5%; 1: <1% of total number of plants present.

of 0.8m being recorded on one occasion. In the three ponds at all seasons the water temperatures follows the ambient air temperatures, and there is little evidence of temperature stratification. All of the ponds are acidic with lower values (more acidic) being found during the summer and autumn months.(5.0-5.9) The surface waters appear to be well-oxygenated but the oxygen content can drop off in the deeper parts of the Rocket Pole and Quarry water bodies in the summer and autumn seasons.

Flora

Ten different plant species were found with the two rushes, *Juncus effusus* and *Eleocharis palustris* occurring at the margins of all three ponds (Table 6.) Both species have spread further around the ponds since 1979. The compact rush, *Juncus conglomeratus*, which also occurs at Pondsbury, was noted for the first time in Rocket Pole Pond in the autumn of 2003. Floating and submerged plants only occurred in the shallow regions of these ponds, e.g. *Hydrocotyle vulgaris*, marsh pennywort, in Rocket Pole and Quarter Wall ponds and *Myosotis scorpioides*, water forget-me-not, near the muddy outflow at Quarry Pool and in the shallow 'beach areas' at the Quarter Wall and Rocket Pole ponds.

Plants apart from *Juncus effusus* were scarce in January 2006; small shoots of *Eleocharis palustris* were appearing and very small plants of *Lythrum portula* (in Quarter Wall pond) and *Myosotis scorpioides* (in all three ponds) were present.

Plankton

Table 7 shows the organisms found in the plankton and indicates the species showing high abundance during the various seasons. The Rocket Pole Pond is eutrophic and experiences algal blooms with different species dominating at different times of the year. The green alga, *Botryococcus braunii*, which appears reddish-brown due to the presence of oil droplets, coloured the water in the spring and in the summer the water appeared 'soupy green' due to blooms of species of blue-green algae, *Microcystis* sp., *Arthrospira* sp. In the autumn the green alga *Desmodesmus magnus* dominated the phytoplankton. No blooms were evident in January 2006, but 'green soupy' water was noted in the Rocket Pole Pond in January 1996 (Richardson *et al.*, 1998). Phytoplankton is less diverse in the other two ponds, but diatoms can build up to fairly large populations in the Quarry Pool in the spring.

In the zooplankton, the Cladocera (water fleas) occur in all of the ponds with *Daphnia obtusa*, *Bosmina longirostris* and *Chydorus sphaericus* present. However *Bosmina longirostris* dominates at all seasons in the Rocket Pole Pond, but appears mainly in the spring in the Quarry and Quarter Wall water bodies. Quarry Pool has the most diverse number of species of rotifers with four species of *Keratella* occurring. The carnivorous rotifer, *Asplanchna priodonta* which feeds on *Keratella* and other smaller rotifers was present in fairly large numbers (abundance rating 3) in Quarry Pool in the autumn. All stages, adults, immature cyclopids, nauplii larvae, of the copepod *Cyclops* were present in all three ponds throughout the year. The larva of the phantom midge, *Chaoborus crystallinus*, which occurred in great abundance in the autumn in Pondsbury also appeared in much smaller numbers in the Quarter Wall pond.

Table 7: Organisms in the plankton of the Rocket Pole Pond, Quarry Pool and the Quarter Wall pond

ROCKET POLE POND	QUARRY POOL	QUARTER WALL POND
CYANOPHYTA (blue-green algae) *Microcystis* sp.*S *Arthrospira* sp.*S *Gomphosphaeria* sp. CHLOROPHYTA (green algae) *Botryococcus braunei* Kutzing *Sp *Pediastrum boryanum* (Turpin) *Desmodesmus magnus* Chodat*A *Ankistrodesmus* sp. Filamentous green alga ZOOPLANKTON CLADOCERA *Daphnia obtusa* Kurz *Bosmina longirostris* (Muller)*A*W *Chydorus sphaericus* Muller *Simocephalus vetulus* (Muller) ROTIFERA *Brachionus calcyflorus* Pallas *Brachionus rubens*(Ehrb.) *Keratella quadrata* (Muller) *Keratella cochlearis*(Gosse) *Keratella vulga* (Ehrb.) *Filinia longiseta* (Ehrb.) CRUSTACEA *Cyclops* sp. Immature cyclopids Cyclopid nauplii Harpacticoids INSECTA Chironominae larvae	CHLOROPHYTA *Pediastrum boryanum*W *Dictyosphaerium* sp. Filamentous green alga BACILLARIOPHYTA (Diatoms) *Pinnularia* sp.*Sp *Tabellaria* sp ZOOPLANKTON CLADOCERA *Daphnia obtusa* *Bosmina longirostris* *Chydorus sphaericus* *Simocephalus vetulus* ROTIFERA *Brachionus rubens* *Keratella quadrata* *Keratella cochlearis* *Keratella serrulata*(Ehrb)*A *Keratella vulga* *Filimia longiseta* *Euchlanis dilatata* Ehrb. *Asplanchna priodonta* Gosse CRUSTACEA *Cyclops* sp. Immature cyclopids Cyclopid nauplii Harpacticoids INSECTA Chironominae larvae	CHLOROPHYTA *Closterium* sp. *Pediastrum boryanum* Filamentous green alga ZOOPLANKTON CLADOCERA *Daphnia obtusa* *Bosmina longirostris* *Chydorus sphaericus* ROTIFERA *Brachionus calcyflorus* *Brachionus rubens*W *Keratella quadrata*Sp*W *Keratella vulga* *Filinia longiseta*W *Polyarthra minor* Voigt CRUSTACEA *Cyclops* sp. Immature cyclopids*S Cyclopid nauplii*S Harpacticoids Cypridid ostracod INSECTA *Chaoborus crystallinus* (Deg.) larva Chironominae larvae

Key: *Sp., *S, *A, *W denote high abundance 5 or 4 in spring, summer, autumn and winter.

Table 8: Abundance of macroinvertebrate species in Rocket Pole Pond, Quarry Pool and Quarter Wall pond

SPECIES	ROCKET POLE POND	QUARRY POOL	QUARTER WALL POND
Oligochaeta:			
Lumbriculus variegatus (Muller)	S1 A2	-	Sp2 S2 A2
Hirudinea:			
Helobdella stagnalis (L.)	-	-	A1 W1
Glossiphonia complanata (L.)	-	-	W1
Crustacea:			
Asellus meridianus Racovitza	S1	Sp2 S2 A2 W1	Sp2 S2 A3 W2
Insecta: Ephemeroptera:			
Cloeon dipterum (L.)	S1	W1	A1
Insecta: Odonata:			
Ischnura elegans (van de Linden)	Sp2 S2 A2 W1	Sp1 S1 A1	A2 W1
Sympetrum striolatum (Charp.)	-	S1	-
Insecta: Hemiptera:			
Gerris gibbifer Schum.	-	S2 A3	-
Notonecta marmorea viridis Delcourt	-	-	Sp2 S2 A2
Immature notonectids	-	-	S2
Corixa punctata (Illiger)	-	-	Sp2 S2 A2
Corixa panzeri (Fieb.)	-	-	S3
Callicorixa praeusta (Fieb.)	-	-	S2
Sigara dorsalis (Leach)	S1	A1	Sp2 S3 A2 W2
Immature corixids	-	S1	Sp2 S3 A2 W2
Insecta: Trichoptera:			
Limnephilus vittatus (Fab.) larva	-	-	W1
Insecta: Coleoptera:			
Ilybius quadriguttatus L. adult	-	S1	S2
Dytiscid Colymbetinae larva	-	S2	S2
Gyrinus substriatus Stephens	-	S2	-
Insecta: Diptera:			
Chironominae larva and pupa	Sp2 S2 A2 W2	Sp1 S1 A2 W1	Sp2 S2 A2 W2
Chaoborus crystallinus (Deg.) l.	-	-	A2
Culex sp. larva	Sp2	-	-
Mollusca:			
Pisidium personatum (Malm)	-	A1	-
TOTAL SPECIES FOUND 21	7	12	18

Key: Sp=Spring, S=Summer, A=Autumn, W=Winter. Abundance scale: 5: >500; 4: 200-499; 3: 50-199; 2: 5-49, 1: <5 individuals.

Macroinvertebrates

Macroinvertebrates found in the three ponds are listed in Table 8, together with an indication of seasonal abundance. The apparently greater species diversity in the Quarter Wall pond is due to members of the water boatman group (notonectid and corixid Hemiptera), which dominate the fauna. Two species of leech, *Helobdella stagnalis* and *Glossiphonia complanata* also occur in this pond. The sheltered waters of the Quarry Pool provide a very suitable habitat for the surface-dwelling species, *Gerris gibbifer*, which was present in the summer and autumn. This species overwinters in cracks and crevices in the steep quarry walls. Five species are found in all three ponds but in small numbers, the isopod *Asellus meridianus* which is so abundant in Pondsbury, the damselfly larva, *Ischnura elegans*, the mayfly larva, *Cloeon dipterum*, the lesser water boatman, *Sigara dorsalis* and a larva of the Chironominae group. The Rocket Pole Pond has fewer macroinvertebrates in both species and numbers than in the other two ponds. This is probably due to the recurring blooms of algae that occur throughout the year and the fairly large population of the mirror carp, *Cyprinus carpio* that is present in this pond.

Fish

No detailed investigations of the fish which are known to occur in two of the ponds have been carried out. Golden carp, *Carassius auratus* are often observed in Quarry Pool and crucian carp, *Carassius carassius* have also been recorded in this pond. A large population of mirror carp, *Cyprinus carpio* exists in the Rocket Pole Pond (George, 1982) but it is difficult to see how this pond with its sparse plant and animal life can support these fish. Feeding by visitors to the island is a contributory factor but cannibalism by the larger fish is highly probable.

DISCUSSION

The freshwater ecosystems on Lundy are governed by the weather with several drying up during periods of drought and many temporary water bodies appearing during periods of intensive rainfall. Their water chemistry relates to the geology of the island which is composed of Tertiary granite. Although there have been surveys of specific groups of freshwater organisms it was not until the late 1970s that detailed investigations were carried out on the flora and fauna of the entire water bodies.

Pondsbury

Dredging of Pondsbury in 1993 and 1995 and the construction of the impounding wall on the western side has affected the surface area of the water body, which in the winter of 2006 had an area of $4000m^2$. The extent of open water has increased since 1979 when it was just 20% of the total area with beds of marsh St John's Wort, *Hypericum elodes*, dominating much of the water body (George & Stone, 1980). In the summer of 1993, before the dredging occurred, Clabburn found that there was 92% open water (Clabburn, 1994). This has been the situation since the dredging with a slightly greater open water area being recorded in January 2006 which is to be expected as many of the plants had died down during the winter months.

The deepest part of Pondsbury is in the northern part where depths of over 1m have been recorded since 1979 at all seasons. The maximum depth in January 2006 was 1.8m. It becomes progressively shallower towards the east and south where depths of a few cm are usually recorded. Water temperatures relate to ambient air temperatures. During the daytime in the summer the surface water temperatures usually increase and there is evidence of thermal stratification in the body of water. Similarly dissolved oxygen levels may fall in the deeper regions but Clabburn in his 24h sampling programme found that in some regions greater levels of oxygen were found at night in the deeper layers.(Clabburn, 1994) In small fairly shallow water bodies there may be overturn of water at night brought about by surface cooling and by the wind. If this occurs oxygen-rich water from the surface layers will reach the deeper regions.

Pondsbury can be classified as a 'soft' water body (mean total hardness 9.4mg/l) with an acid pH that is maintained by the extensive growth of *Sphagnum* moss that dominates the area. *Sphagnum* has the ability to bind cations and release hydrogen ions in their place thus maintaining acidity. pH values of 4.8 have been recorded in the summer surveys, but in the other seasons the pH appears to increase (become less acid). The pH can vary in acidic waters which are poorly buffered due to changes in the free carbon dioxide content. The presence of plants and animals in the water can affect CO_2 levels by their respiration which increases CO_2 content in the water. The fewer plants and animals in the water in the colder months of the year produce a smaller amount of free CO_2 in the water which allows the pH to rise, becoming less acid. pH values of 6.4 were recorded in the winter of 2006.

The composition of the **flora** reflects the acidic nature of the water with the dominant plants being characteristic of bogs and marshes (Table 2). Although the species composition has remained remarkably stable over the last 27 years, the relative abundance of most species has varied considerably. *Sphagnum cuspidatum*, the bog moss, still dominates Pondsbury and the surrounding boggy area, but the other dominant, the soft rush, *Juncus effusus* has increased during the last 27 years, particularly around the margins of the north and south side. The spike rush, *Eleocharis palustris*, dies back during the winter but the spring 2003 survey showed that it had increased particularly on the eastern side since the earlier surveys. The large 'island stands' of *Hypericum elodes*, which were so noticeable in the summers of 1979 and 1986 no longer exist, and only small patches are present, mainly on the southern side. The 1993 and 1995 dredging (Gibson 1994; Parkes 1996 & R. Lovell pers. comm.) are obvious reasons for its decline. This decline however may have favoured the growth of *Potamogeton polygonifolius*, the bog pond weed, which has become more abundant in the last 10 years.

The **plankton** population shows good species diversity with many of the Crustacea and Rotifera recorded by Galliford in 1953 still present (Galliford, 1954). The composition of both the phytoplankton and the zooplankton varies throughout the year with often one species dominating for a short period. Although species may increase in numbers at one particular season they are usually present throughout the year in small residual populations (Moss, 1980). The two well represented groups,

Cladocera (water fleas) and Rotifera (rotifers) have resting egg stages that *can* withstand adverse conditions. *Daphnia obtusa* was a prominent species throughout the year, but its smaller numbers in the autumn were probably due to the dominating presence of the predatory larva of the phantom midge, *Chaoborus crystallinus* which formed about 50% of the total plankton numbers at that time (Figure 1). The *Keratella* rotifers are represented by three species, *K. vulga*, *K. quadrata* and *K. serrulata*, with the latter being very common in *Sphagnum* bogs and acid waters on the mainland. Both the cyclopoid and harpacticoid groups of copepods occur in Pondsbury with all life cycles of *Cyclops* sp appearing throughout the year. The other free-living group of copepods which is found on the mainland, the Calanoida, appears to be absent from Lundy. Green algae were present in Pondsbury throughout the year with different species appearing at different seasons. The dominance of green algae (Figure 1) during the summer months was mainly due to a 'bloom' of the green Euglenophyte, *Euglena viridis*.

Although there are seasonal differences in the composition and numbers of **macroinvertebrates** in Pondsbury, the fauna appears to show marked similarity to that observed 27 years ago. In spite of the dredging and the decline of *Hypericum elodes*, the isopod crustacean, *Asellus meridianus* remains the dominant member of the fauna, with the flatworm, *Polycelis nigra* and members of the Coleoptera, Hemiptera and Chironominae again well represented groups. The more common mainland form *Asellus aquaticus* does not occur on Lundy and this is in agreement with the findings of Williams (1962, 1979) and Moon & Harding (1982) who found only *Asellus meridianus* on offshore islands. *Polycelis nigra*, the black flatworm, although present all year round, was particularly abundant in the cold winter waters. The acid water and the fairly high summer water temperatures exclude some species, particularly members of the Ephemeroptera, mayfly larvae, which are much more abundant on the mainland. Only one species, *Cloeon dipterum* has been found. A long-standing macroinvertebrate, the water spider, *Argyroneta aquatica*, which lives amongst the *Sphagnum* is still at Pondsbury although in smaller numbers than in 1979 when fairly large numbers were recorded. This spider was recorded as 'quite abundant' at Pondsbury 53 years ago (Galliford, 1954). The drying up of the water body presents another hazard to macroinvertebrates particularly those that remain in the water all the time. Coleoptera and Hemiptera can fly away to other waters if conditions deteriorate. Other species survive by the formation of resistant cysts/cocoons (leeches, oligochaete worms, flatworms) and others can aestivate in the bottom sediments. The isolation of Lundy from the mainland may deter some macroinvertebrates from reaching the island, but it is more likely to be the water conditions that determine the composition of the fauna. There appear to be no endemic species in the freshwater flora and fauna.

Rocket Pole Pond, Quarry Pool, Quarter Wall pond
These three ponds which have been formed from excavations in the rock, are of similar size. They maintain their shape throughout the year although the water levels fluctuate according to weather conditions. Rocket Pole Pond is a very

exposed body of water, subject to strong winds in all seasons. Its water is well mixed and if temperature and oxygen stratification does occur during warm periods, it will quickly be overturned. The shallow Quarter Wall pond is also fairly well exposed particularly to easterly winds. The water appears fairly uniform in its temperature and oxygen content throughout its depth. Quarry Pool is a sheltered body of water particularly from the westerlies which frequently blow across the island. There is some evidence of a decrease in oxygen content in the deeper waters in the summer which may be due to decomposition of organic matter which is prevalent in the bottom sediments.

The acidic nature of all three ponds determines the type of flora and fauna present. The number of species of **plants** has remained remarkably consistent since surveys began in 1979. Due to the depth of the ponds, plants mainly occur around the margins and one in particular, *Juncus effusus* is fairly abundant at all three ponds. The spike rush, *Eleocharis palustris* has increased in recent years at all of the ponds The few other plants, all characteristic of acidic upland waters, occur in the shallow regions.

The **plankton** samples taken during all seasons show that the Rocket Pole Pond is eutrophic with algal blooms regularly occurring. There is no through flow in this pond and there is a build up of nutrients from the mirror carp population and the droppings from ducks that regularly frequent this pond. The summer blue-green algae bloom was also noted by Galliford in 1953 (Galliford, 1954). Evidence of eutrophy in the summer is also seen in the Quarter Wall pond which is used by the ponies, where large populations of green algae, particularly the desmid, *Closterium* and filamentous green algae often occur. Quarry Pool does not appear to experience algal blooms but large populations of some species occur at certain times of the year e.g. the diatom *Pinnularia* in the spring and the green alga *Pediastrum boryanum* in the winter.

Differences can be observed in the composition of the zooplankton in the three ponds. Although the cladoceran, *Bosmina longirostris* is found in all of them, it is much more prolific in Rocket Pole Pond where it reaches large populations in autumn and winter (abundance rating 5). *Daphnia obtusa* which was the main cladoceran in Pondsbury also occurs in these water bodies, particularly in the winter and spring of Rocket Pole Pond and Quarry Pool and in the summer at the Quarter Wall pond, where it is often found with the rotifer, *Brachionus rubens* attached to it. This commensal relationship was recorded also in this pond by Galliford in 1953. Several of the female *Daphnia* in the Quarter Wall pond autumn samples were carrying the overwintering resting eggs. All life cycle stages of the copepod *Cyclops* occurred in the ponds. This species is known to breed throughout the year and it is not unusual to find nauplius larvae and immature forms as well as egg-carrying females at all seasons of the year (Harding & Smith, 1974). Harpacticoid copepods were present but in much smaller numbers, but as in Pondsbury calanoid copepods, which on the mainland are more frequent in the winter, were not found. 10 species of rotifers were present in the ponds with greater species diversity in the Quarry Pool. Here the rotifer, *Keratella serrulata* dominated the autumn plankton (68% of the total population).

Several of the cladoceran and rotifer species were recorded by Galliford 53 years ago, and appear therefore to be long standing members of the plankton e.g. *Daphnia obtusa, Bosmina longirostris, Chydorus sphaericus, Keratella serrulata, Keratella vulga, Brachionus rubens*. The resting egg stage which is found in both groups together with their parthenogenetic life cycle which ensures the fast build up of large populations, contribute to the success of these species in the Lundy lentic waters.

Differences are observed in the **macroinvertebrate** populations of the three ponds, and these can be related to the position of the pond on the island and hence exposure to the elements, the amount of plants growing in these ponds and the nutrient input. The algal blooms that occur so frequently in the Rocket Pole Pond for most of the year probably deter many macroinvertebrates and there are no large populations of any species at any time of year. A contributory factor to the low species diversity and numbers of organisms is the presence of the mirror carp, *Cyprinus carpio*, which feed on algae and invertebrates. Although more species were found in the Quarry Pool only one species, the pond skater, *Gerris gibbifer* reached reasonable numbers (abundance rating 3). Fish, golden carp, *Carassius auratus* and crucian carp, *Carassius carassius* are also present in this pool and it is likely that their predation on macroinvertebrates partly attributes to the low numbers of organisms found. This was demonstrated by Macan (1966) who studied the effects of fish predation on the fauna of upland ponds. The sheltered nature of Quarry Pool allows the surface-dwelling *Gerris gibbifer* and the whirligig beetle, *Gyrinus substriatus* which was found in the summer to live successfully here. *Gerris* in particular builds up fairly large populations and both adult and young forms occurred in the summer and autumn.

The shallower Quarter Wall pond has more plant beds e.g. *Myosotis scorpioides* and *Lythrum portula* than the other two ponds, and this explains the presence of the water boatman group of the Hemiptera, particularly the lesser water boatmen, corixids and sigarids, which are plant feeders. The greater water boatman, *Notonecta marmorea viridis*, which also occurred (abundance rating 2) although preferring stretches of open water, places its eggs in plant stems during the spring.

Seasonal differences were observed in the three ponds with fewer species being found during the winter months. Water temperatures were very low in January 2006 at the time of sampling. More species were recorded at the Quarter Wall pond where two species of leech, *Glossiphonia complanata* and *Helobdella stagnalis* were found amongst the emerging *Myosotis* beds, as well as *Asellus meridianus, Sigara dorsalis* and immature corixids. It is unlikely that the fairly deep Rocket Pole Pond and Quarry Pool dry up during long periods of drought although water levels can drop. The Quarter Wall pond however has suffered severe water loss in some years e.g. in the summers of 1981, 1995 and 2006, but the dredging of part of the pond in the autumn of 1995 (Parkes, 1996 and R. Lovell pers. comm.) has helped its survival in the past 10 years. Many of the animal species can survive periods of desiccation by the production of resting eggs (Cladocera, Rotifera), as resistant cysts (Oligochaeta, Platyhelminthes), aestivation in the bottom sediments

(*Asellus meridianus*, insect larvae) or by flying off to more permanent water bodies (adult Coleoptera, Hemiptera).

TEMPORARY WATERS
During the five surveys since 1979 some of the smaller temporary waters were studied and descriptions of their flora and fauna have been given in previous papers (George & Stone, 1981; George & Sheridan, 1987; George, McHardy & George, 2004). Brief descriptions of the water bodies are given here.

Small pond at Quarter Wall (OS grid reference: SS 13630 44965)
The smaller pond at Quarter Wall is shallow with a dense weed cover and no open water. It is situated in a depression (3m x 6m) in a marshy area where there are stands of the soft rush, *Juncus effusus* and there is a small outlet on the eastern side. Its depth varies according to weather conditions and the maximum depth recorded during the surveys was 0.3m. In dry periods e.g. summer of 1995 and summer/ autumn 2003, it dries up altogether. The water temperature follows the ambient air temperature and usually there is abundant oxygen due to the prolific growth of plants. The pH varies from 5.0 in summer to 6.2 in spring. Clumps of *Juncus effusus* surround the pond and the dominant plant in the pond is the bog pondweed, *Potamogeton polygonifolius* which is present all year round. In the spring *Myosotis scorpioides, Callitriche stagnalis* and *Lythrum portula* appear and in the summer *Hydrocotyle vulgaris* has been found.

The pond supports a good population of *Asellus meridianus*, which feeds on the large amounts of decaying vegetation that are present. Hygrobatid mites are common in the summer and several insects have been found e.g. the damselfly larva, *Ischnura elegans*, mayfly larva, *Cloeon dipterum*, various corixids and water beetles. It is interesting to note that the water beetle, *Helophorus grandis* which prefers to crawl amongst plants and does not swim, occurs in this small pond.

Pools in the North Quarry (OS Grid Reference: SS 138833 45597)
The two pools in the North Quarry are shallow and covered with aquatic plants. The maximum depth in the pool nearest to the quarry entrance on the south side was 0.73m in October 2003 and 0.57m in April 2005. The pool adjacent to the steep quarry wall on the south side and completely surrounded by large rocks, had a maximum depth of 0.9m in October 2003 and 1.25m in April 2005. pH values for both varied from 5.0 in the autumn to 5.35 in the spring. The greatest diversity of plants occur in the shallower pool where seven species were recorded in April 2003. Three species were found in the deeper pool. Small stands of *Juncus effusus* occur in both pools, but *Callitriche stagnalis* dominated forming a green carpet across both water bodies. *Sphagnum cuspidatum* occurs in the shallow marginal areas. The shallower pool also has, *Myosotis scorpioides*, the water forget-me-not, *Ranunculus flammula*, the lesser spearwort, *Caldesia parnassifolia*, water plantain and the duckweed, *Lemna minor*. In the 2006 winter both pools were covered with *Lemna minor* and in the shallower pool small shoots of *Myosotis* were emerging.

The dense weed cover in both pools provides good shelter for several macroinvertebrates, particularly *Asellus meridianus* and aquatic beetles. A caddis larva, *Plectrocnemia conspersa* which was not found in the permanent ponds, occurred in the deeper pool. This larva which spins a net to catch its prey, usually emerging insects, has been found in the Lundy streams (Long, 1994). It has been recorded living in upland pools and lakes as well as rivers and streams (Edington & Hildrew, 1995).

David's Pool (OS Grid Reference: SS 13846 44228)

This small pool is situated adjacent to the Pondsbury raised dam on the western side. Maximum depths in Autumn 2003 and Spring 2005 were 0.6m and 0.66m respectively. As expected the same seasonal pH values as Pondsbury were recorded, *Callitriche stagnalis*, the mud water starwort, dominated the pool throughout the year, with stands of *Juncus effusus* surrounding and encroaching into the water body. As the water temperatures increased, patches of *Hydrocotyle vulgaris*, and *Potamogeton polygonifolius* appeared with the water crowfoot, *Ranunculus omiophyllus* appearing later in the season.

Several of the animals found in Pondsbury occurred in this pool as expected, and they included several of the planktonic organisms, e.g. *Daphnia obtusa*, *Bosmina longirostris*, *Cyclops* sp. The black flatworm, *Polycelis nigra*, *Asellus meridianus*, various beetles and chironomid larvae were present.

Ray's Pool (OS Grid Reference SS 13486 44228)

For the first time in April 2005 the pool (named Ray's Pool after the farmer's sheepdog) at the bottom of Barton Cottages field was surveyed and again visited in January 2006. This spring-fed pool had a maximum depth of 0.75m and the pH varied from 6.0 in the spring to 6.52 in the winter. In both seasons it was covered with the water cress, *Nasturtium officinale* and *Juncus effusus* surrounded the banks on the north and south sides. *Ranunculus omiophyllus* occurred in the spring in the shallower areas neat the outlet on the eastern side but was not evident in January 2006.

The most dominant macroinvertebrate present at all seasons is the gastropod mollusc, *Lymnaea peregra*. This mollusc, one of the commonest in Europe is not found in other Lundy lentic waters, where only a few molluscs have been recorded. Molluscs prefer more alkaline calcareous waters, but some species such as *Lymnaea peregra* can tolerate soft acid waters. A total of over 200 *Lymnaea* were collected in two one-minute net sweeps through the water cress beds. Other macroinvertebrates found were the ubiquitous *Asellus meridianus*, the mayfly larva, *Cloeon dipterum*, corixids and chironomid larvae. The small beetle *Laccophilus minutus* was fairly abundant.

Johnny's Pool (OS grid Reference SS 132879 47237)

A small pool, 4.8 m long and 1 m wide, on the west side of the main track at Gannets Combe was covered with the floating club-rush, *Eleogiton fluitans* in April 2005 and January 2006. This is a perennial species typical of shallow acidic waters and it frequently grows in dense masses forming a bright green carpet throughout the year. The pool which has a depth range of 0.2-0.9m has a pH of 6.0. *Ranunculus omiophyllus* and *Lythrum portula* also occur. Adults and larvae of the Hydroporinae beetle sub-family were found amongst the *Eleogiton* and the oligochaete worm *Lumbriculus variegatus* was also recorded.

Rocket Pole Temporary pond (OS Grid Reference SS 13481 43681)
The large depression to the east of the Rocket Pole contains water at various times of the year. In January 2006 it contained dense mats of *Eleogiton fluitans* and filamentous green algae. A pH of 6.3 was recorded at this time and this is in accord with the January pH values recorded in ponds in the Rocket Pole area in January 1966. (Richardson *et al.*, 1998). Microscopic Crustacea, such as *Daphnia obtusa* and ostracods were abundant in the water body.

CONCLUSIONS
The flora and fauna of the Lundy lentic waters although typical of acidic waters on the mainland, are impoverished compared with them, but the isolation of Lundy is probably not a major limiting factor. There are no endemic species or varieties present.

The four main ponds display differences particularly in their plankton and macroinvertebrate communities and these can be related to the position of the water body on the island and hence exposure to the elements, the amount of vegetation present and their nutrient content. Pondsbury, the largest water body has the greatest species diversity and numbers of organisms. The Rocket Pole Pond frequently experiences algal blooms which markedly affect the macroinvertebrate populations.

The seasonal surveys of the four main ponds have shown that plankton is present throughout the year but with different species dominating at different times. Many of the plants die down during the winter months and the macroinvertebrates decline in numbers. The temporary pools which frequently occur on the island quickly become colonized and their communities, particularly the plants and some macroinvertebrates appear to survive periods of desiccation.

The flora and fauna of Lundy's lentic freshwaters has shown a remarkable stability in the species composition over the last 27 years since the main surveys began in 1979. Some of the organisms which were recorded by early field workers over 50 years ago are still present such as *Asellus meridianus*, and the water spider, *Argyroneta aquatica* It is the isolation of these waters on the island and little human interference that has contributed to the long term stability of these ecosystems.

Pictures of the ponds taken in various years and seasons are given in Plates 1-6, pages 126-128.

ACKNOWLEDGEMENTS
The author wishes to thank the following organisations for funding which has contributed to this work: the Lundy Field Society (1979, 2003), the World Wide Fund for Nature (1979), the University of Westminster (1986, 2003). The author is indebted to her fellow co-worker, Brenda McHardy (Stone) who carried out the plant

surveys in 1979 and 2003. The recent field surveys (2003, 2005, 2006) would not have been possible without the capable assistance of David George and John Hedger. Graham Coleman arranged the loan of field equipment and laboratory facilities for the plankton examination from the University of Westminster, and assisted with the field work in January 2006.

REFERENCES

Alexander, K.N.A. 1992. The Hemiptera of Lundy. *Annual Report of the Lundy Field Society 1991,* 42, 101-105.

Baillie, C.C. & Rogers, M. & W. 1997. Sizes and ages of some crucian carp on Lundy. *Annual Report of the Lundy Field Society 1976,* 27.65-66.

Brendell, M. 1976. Coleoptera of Lundy. *Annual Report of the Lundy Field Society 1975,* 26, 29-53.

Clabburn, P.A.T. 1994. Freshwater biological survey of Lundy, 1993: further studies of the fauna of Pondsbury. *Annual Report of the Lundy Field Society 1993,* 44, 73-83.

Edington, J.M. & Hildrew, A. 1995. *A Revised Key to the Caseless Caddis Larvae of the British Isles.* Freshwater Biological Association Scientific Publication, no. 53. Cumbria: Freshwater Biological Association.

Fraser Bastow, R. 1950. Freshwater diatom flora. *Annual Report of the Lundy Field Society 1949,* 3, 32-41.

Galliford, A.L. 1954. Notes on the freshwater organisms of Lundy with especial reference to the Crustacea and Rotifera. *Annual Report of the Lundy Field Society 1953,* 7, 29-35.

George, J.J. 1982. The mirror carp, *Cyprinus carpio,* of the Rocket Pole Pond. *Annual Report of the Lundy Field Society 1981,* 32, 38-39.

George, J.J. 1997. The freshwater habitats of Lundy. In R.A. Irving, A.J. Schofield & C.J. Webster (eds), *Island Studies. Fifty Years of the Lundy Field Society,* 149-164. Bideford: Lundy Field Society.

George, J.J., McHardy (Stone), B.M. & George, J.D. 2004. Further investigations of the flora and fauna of the Lundy lentic freshwaters. *Annual Report of the Lundy Field Society 2003,* 53, 110-130.

George, J.J., McHardy (Stone), B.M. & Hedger, J.N. 2004. A comparative investigation of the plankton of the four permanent Lundy lentic freshwaters. *Annual Report of the Lundy Field Society 2003,* 53, 99-109.

George, J.J. & Sheridan, S.P. 1987. Further investigations of the flora and fauna of the freshwater habitats. *Annual Report of the Lundy Field Society 1986,* 37, 35-46.

George, J.J. & Stone, B.M. 1980. The flora and fauna of Pondsbury. *Annual Report of the Lundy Field Society 1979,* 30, 20-31.

George, J.J. & Stone, B.M. 1981. A comparative investigation of the freshwater flora and fauna of the Lundy ponds. *Annual Report of the Lundy Field Society 1980,* 31, 19-34.

Gibson, A. 1994. Wardens's report 1993. *Annual Report of the Lundy Field Society 1993,* 44, 7-9.

Harding, J.P. & Smith, W.A. 1974. *A key to the British Freshwater Cyclopid and Calanoid Copepods*. Freshwater Biological Association Scientific Publication, no. 18. Cumbria: Freshwater Biological Association.

Lane, R.P. 1978. The Diptera (two-winged flies) of Lundy Island. *Annual Report of the Lundy Field Society 1977*, 28, 15-31.

Langham, A.F. 1969. Water courses and reservoirs on Lundy. *Annual Report of the Lundy Field Society 1968*, 19, 36-39.

Long, P.S. 1994. A study into the macroinvertebrate fauna and water quality of Lundy Island's lotic environment. *Annual Report of the Lundy Field Society 1993*, 44, 59-72.

Macan, T.T. 1966. The influence of predation on the fauna of a moorland fishpond. *Archiv. Hydrobiol.*, 66, 432-452.

Moon, H.P. & Harding, P.T. 1982. The occurrence of *Asellus* (Crustacea, Isopoda) on offshore islands in the British Isles. *The Naturalist*, 107, 67-68.

Morgan, H.G. 1948. Aquatic habitats. *Annual Report of the Lundy Field Society 1947*, 1, 12-13.

Moss, B.V. 1980. *Ecology of Freshwaters*. Oxford: Blackwell Scientific Publications.

Parkes, E. 1996. Warden's report for 1995. *Annual Report of the Lundy Field Society 1995*, 46, 6-7.

Richardson, S.J., Compton, S.G. & Whiteley, G.M. 1998. Run-off fertiliser nitrate on Lundy and its potential ecological consequences. *Annual Report of the Lundy Field Society 1997*, 48, 94-102.

Williams, W.D. 1962. The geographical distribution of the isopods, *Asellus aquaticus* (L.) and *Asellus meridianus* Rac. *Proceedings of the Zoological Society of London*, 139, 75-96.

Williams, W.D. 1979. The distribution of *Asellus aquaticus* and *A meridianus* (Crustacea, Isopoda) in Britain. *Freshwater Biology*, 9, 491-501.

Plate 1: Pondsbury in October 2003. *(Photo: David George)*

Plate 2: Pondsbury in August 1979 showing the extensive beds of Marsh St John's Wort, *Hypericum elodes. (Photo: David George)*

Plate 3: The Rocket Pole Pond in April 2005 with St Helena's Church in the background. *(Photo: David George)*

Plate 4: Depth measurements being taken at Quarry Pool in October 2003. *(Photo: Jennifer George)*

Plate 5: The larger pond at Quarterwall in April 2005. *(Photo: David George)*

Plate 6: The larger pond at Quarterwall in August 2006 showing the effects
of the summer drought. In September it dried up completely.
(Photo: Alan Rowland)

MARINE AND FRESHWATER ECOLOGY: DISCUSSION

Marine: *(Initials: KH=Keith Hiscock, RI=Robert Irving, Q=Unknown participant)*

Q: How did the Mediterranean corals reach Lundy - by currents or by dropping off the bottom of a boat?

KH: We mentioned the decline in some of the marine species that is occurring in Lundy waters. There are long term cycles in the marine environment e.g. the Russell Cycle. Every 30/40 years there are periods of about five years when the abundance of larval fish and larvae of benthic invertebrates is very high. During the 1950s and 1960s the North Atlantic Oscillation was in a strong negative phase but is now becoming positive. It is the character of water masses that creates good conditions for some species to recruit well and survive, but these conditions have not been good for some time.

To answer your specific question: the Devonshire cup coral has a fairly long planktonic larval stage, up to a month, and it can settle considerable distances from the adults and consequently the species recruits readily. The Mediterranean yellow cup corals have a short larval phase and usually settle close to the adults. These corals occur in five known areas in Britain - Lundy, Lyme Bay, Plymouth, Scilly Isles and off the Lizard peninsula. These populations can either be relicts of a much larger widely-distributed population or can be brought over from Brittany by jet-stream currents and deposited by chance at Lundy and in the other areas. They need a strong jet-stream current to get over from Brittany and these currents do occur. They were discovered at the end of the eighteenth century. It is important to remember the long term changes I mentioned - every several decades there are very good conditions for species recruitment and survival. I am sure that the Lundy species will increase in the future.

Q: In freshwaters there is often a more restricted flora and fauna with speciation occurring. This is not the case in the marine environment. Also why do you find such a great diversity of plants and animal species in the waters around Lundy?

RI: There is far less discontinuity between different marine habitats as all are linked by the sea. Lundy has a variety of different habitats - exposed, sheltered, strong tidal stream areas, which allows a diversity of plants and animals to live, compared with nearby coastal waters. There is not the speciation and endemism occurring as in the isolated freshwater habitats.

KH: The importance of Lundy in terms of its marine life is, as Robert says, its great variety of habitats, shallow, wave-exposed, sheltered, rock, sediments; it is this variety that allows such a great diversity to develop and survive.

Q: Has there been an evaluation of the No-Take Zone with regard to teleost and elasmobranch fish?

KH: Fish surveys were considered in the initial planning of the monitoring programme. Fish-watching stations and the counting of fish are difficult to set up and it is easier to use sedentary animals for quantitative assessment with meaningful statistics. It would be more difficult with the bottom-living dogfish (elasmobranchs), but it is possible with some teleosts e.g. wrasse, as techniques are available. The monitoring of fish should be considered in the future.

Freshwater: *(Initials: JG=Jennifer George, RA=Roger Allen, KG=Keith Gardner, DK=Diana Keast, JM=John Morgan, Q=Unknown participant)*

Q: Have any amphibians been recorded on Lundy?

JG: There are no records of amphibians as far as I know. Amphibians rarely occur on isolated islands as they are not good travellers.

RA: *I was surprised to hear that there are carp in Pondsbury, as it is very shallow. Also there is a large reservoir at the end of The Quarters which used to contain fish. Are they still there?*

JG: In 1976 Pondsbury dried up and Chris Baillie and Mick Rogers transferred Crucian carp from Pondsbury to the Quarry Pool. In January 1977, 30 Crucian carp were returned to Pondsbury. In 1986 we saw several, but a much-needed survey of the fish has never been done. Pondsbury is now 1.8m deep in places, particularly after the dredging in 1995.

KG: There is a medieval document listing two tenements and a vivary. A vivary is a monastic term for fish pond which usually contains carp. There could have been carp on Lundy at that time, but I am not inferring that the present carp have medieval ancestors.

JG: There have been introductions of fish, for example Martin Coles Harman introduced golden carp to Quarry Pool. Perhaps, his daughter, Diana Keast, can help here?

DK: My father did move the fish about from pond to pond quite a bit.

Q: What other fish occur in the Lundy freshwaters?

JG: Tench occur in Quarry Pool as well as Crucian and Golden carp. Rocket Pole Pond mainly has a large Mirror carp population. With regard to the other question about fish in the Quarters reservoir, as far as I know these have never been studied. (N.B. In November 2006, Roger Fursdon informed us that when the Quarter Wall pond dried up in September 2006, he rescued about 100 rudd and placed them in the Rocket Pole Pond).

Q: *You mentioned that the isopod Asellus meridianus is often the only species found on islands. Is it a definite species?*

JG: It is a definite species. On the mainland it does occur with the more common *Asellus aquaticus*, but on islands it is often the only species found, as on Lundy, the Isle of Man and the Scilly Isles.

Q: *How does Asellus get to Lundy?*

JG: The female carries the eggs in a ventral brood pouch and the eggs remain viable for some time. Freshwater organisms can be picked up by birds and carried from one habitat to another and probably this species was picked up from the mainland and transported to Lundy many years ago.

JM: *You mentioned the high level of nitrate in the Rocket Pole Pond. Is this due to the large population of Mirror carp present and the fact that they are fed by visitors in the summer?*

JG: The high levels of nitrate and phosphate are due to the Mirror carp and also the ducks that regularly frequent this pond. Carp feed mainly in the summer and feeding by visitors probably helps their survival. The pond has no through drainage which allows the nutrients to build up, and this is the reason for the prolific algal blooms that frequently occur, giving the pond a 'green soupy' appearance.

THE TERRESTRIAL ECOLOGY OF LUNDY: ORIGINS, PROGRESS AND THE FUTURE

by

TONY PARSONS

Barnfield, Tower Hill Road, Crewkerne, Somerset, TA18 8BJ
e-mail: tony.parsons@care4.complimentary.net

ABSTRACT

The possible origins of the terrestrial fauna and flora of Lundy are discussed. Early records and some of the recorders of the nineteenth century and the first half of the twentieth century are considered. The progress made since 1946 followed two major advances, the publication of *The Fauna and Flora of the Ilfracombe District* and the inauguration of the Lundy Field Society, which stimulated interest in the natural history of the island and led to numerous contributions to its study. The importance of Lundy in conservation terms, the effects of an upsurge in general interest in natural history, the influence of commercial factors, and possible future directions for studies are discussed.

Keywords: *Lundy, terrestrial fauna, terrestrial flora, bird records, conservation*

INTRODUCTION

In the Annual Report for 1967, Keith Gardner (1968) posed the question 'Lundy - a Mesolithic Peninsula?'. Keith concluded that the answer to that question must be 'yes', a conclusion with which I would concur. Following the last Ice Age, there is ample evidence worldwide of post-glacial fluctuations in sea level which is considered to have receded to 100 metres below the present level during the last glaciation (Zeuner, 1950). The island itself was not directly involved in the most recent glaciation, which probably reached the south coast of Wales, but adverse climatic conditions would have restricted the arrival of most species until some time after 10,000 B.C.

As the ice melted, the sea level rose and the re-opening of the Straits of Dover flooded the minus 40 metre level at approximately 7,000 B.C. It is reasonable to assume a comparable date for the Bristol Channel which suggests that Lundy was connected with the mainland for a significant period of time in the post-glacial period, during which the land bridge from Europe was still in existence and the post-glacial climate was increasingly clement.

ORIGINS

The fauna and flora

Immigration using such a connection would have an arbitrary origin for the earliest of the island's fauna and flora of perhaps 8,000 B.C. This would mean that species sensitive to a cooler climate would have been unlikely to reach the island prior to its isolation. The composition and the continued existence of such an immigrant community into historic times would depend on numerous factors, not least the requirement for species to have survived in the proximity of glaciation or to have returned from continental Europe at an early date.

There is one small vertebrate which may have taken advantage of access to Lundy - the pygmy shrew (*Sorex minutus*). This appears to be the only one of the smaller, flightless mammals to have made its way to the island and to have survived to the present day (Bull & Parker, 1997). However, the assumption that it arrived independently and survived for several thousand years leads to the acknowledgement that there must already have been sufficient invertebrates on the island to support a population of these voracious little animals.

By no means all our common flightless species managed to arrive in time for the crossing. Our two commonest grasshoppers in Britain are the meadow grasshopper *(Chorthippus parallelus)* which is predominately flightless and which does not occur on Lundy, and the field grasshopper *(Chorthippus brunneus)* which flies well and which does so occur.

The best candidate for aboriginal immigrant status must be the tiny bagworm moth *Luffia feuchaultella* (Psychidae) which exists only as flightless, parthenogenetic females within small silken cases adorned with lichens. The moth occurs primarily on lichen-covered wood and rocks close to the coast and would be likely to spread at only a few metres a year and may well have survived close to the boundary of glaciation. To the present time, the moth has not been recorded on Lundy but I am sure that it will be found and that this will support the theory of a land-bridge to the island in the Mesolithic period.

There are many other candidates for aboriginal status such as flightless beetles, woodlice, centipedes and millipedes but there are many ways in which these might have arrived. Very large quantities of supplies have been transported to the island over several centuries and it is inevitable that many species will have been introduced even though not all such introductions will have been viable.

The majority of winged species could have arrived at any time but it may be possible to make some assessments of flightless species on the basis of probabilities. For example, the probability of introducing a species which may be synanthropic, such as the common large centipede *Lithobius forficatus*, must be higher than that of introducing a totally 'wild' species such as *Lithobius borealis*.

The origins of recording
The history of the fauna and flora of any site prior to the nineteenth century relates almost entirely to what has been written down in documents such as Court Rolls, diaries and account books where these have survived to the present day. A number of such early documents (referred to in Chanter, 1887 and Langham, 1994) provide minimal details of the terrestrial fauna. For example, the first mentions of rabbits come from 1225 when a dozen were introduced, from an 'Inquisition of Escheat' in 1274, and from further references from 1321 and from the 1580s (Camden, 1607) when the island was said to swarm with rabbits and black rats. Peregrines appeared for the

first time in the 1274 inquisition as did gannets which, in 1321, were noted to breed and, in 1607, the presence in winter of starlings and woodcock was noted. By 1725, deer had been introduced and, by 1787, the 'many deer and goats' were recorded as having been a part of numerous introductions of 'all sorts of game'. In 1775, we are told specifically that there were only black rats on the island, no brown rats, and in 1787 came the first mentions of razorbills, guillemots, puffins and kittiwakes.

The Victorians
The first scientific survey of the island's fauna came from Thomas Wollaston who visited the island in 1844 and 1845 and published his findings in 1845 and 1847. Wollaston was a highly respected gentleman-naturalist of a breed that flourished in the Victorian and Edwardian eras. They were educated, diligent and competent workers in their chosen fields, usually (although not always) comparatively affluent. They were certainly not the idle rich, however. Many of them were professional men who led extremely busy working lives as well as having absorbing hobbies. Despite being only 22 when he first visited Lundy, Wollaston managed to find what were then considered to be 153 different species of beetle, including a species new to science - what we now refer to as *Psylliodes luridipennis*, one of the beetles found on Lundy cabbage and one which is endemic to Lundy.

In 1851, Philip Gosse came to live in Torquay. Already an accomplished author, he was an expert in all aspects of the sea shore. He regularly visited Ilfracombe and, in July 1852, was invited by Hudson Heaven to stay on Lundy for a few days. This sojourn resulted in four chapters in his next book (Gosse, 1865) and his observations of the fauna and flora provide an invaluable background by an accomplished naturalist.

The publication of synopses of existing records is particularly important where early data may be in obscure sources. The first to do this for Lundy was John Roberts Chanter, solicitor and prominent citizen of Barnstaple whose *History of Lundy Island* (Chanter, 1887) is remarkable in that it is believed that Chanter never actually visited the island. Apart from his researches on the earliest relevant data, he repeated existing records and extracted some data from Gosse's book, with additional information in his 1887 edition on the birds, the Lepidoptera and the flowering plants and ferns, these provided by Lundy's owner, the Rev. Hudson Heaven.

Dr George Longstaff was a general practitioner in Wandsworth. He came from a wealthy family and, in 1890 at the age of about 40, he retired from medicine and moved to Mortehoe to concentrate on his great love, the Lepidoptera. He travelled widely and wrote one book on his travels but his other work, from 1907, although on Mortehoe parish, included the Lepidoptera of Lundy (Longstaff, 1907).

PROGRESS
The first half of the twentieth century
Up to this point in time, the data have come from relatively few sources and consist primarily of lists of species although, already, some indications of status have appeared. Another general practitioner, Dr Norman Joy from Reading, who became one of the best known coleopterists of the twentieth century, visited Lundy in 1905

and 1906, on the latter occasion with his friend J.R. le B. Tomlin. They published eight papers and notes between them as a result of their work (Brendell, 1976).

Much of the recording up to this point had been of beetles and even spider expert William Bristowe, who visited Lundy in 1928, recorded a few beetles and other insects as well as publishing the first paper on the spiders of the island (Bristowe, 1929). This was more than a list; for the first time there were considerable annotations regarding locations and status - sufficient in many cases for comparisons with the present day.

Up to the 1930s, studies on the flora had been primarily the list of Hudson Heaven (some records later being considered doubtful) supported by Gosse's notes but yet another general practitioner came to the rescue. Frederick Elliston Wright was the doctor in Braunton, with an abiding interest in natural history and particularly botany, and his papers on the flora and its origins (1934 and 1935), have formed the basis for most studies on these since. Wright's great claim to fame was that he realised that the Lundy cabbage was something unusual. Samples were sent to O.E. Schulz, who described the new species, which is endemic to Lundy, as 'wrightii' in 1936.

In 1939, Richard Perry spent five months on Lundy studying seabirds (Perry, 1940). The book which resulted gave the first extensive information on the seabirds, acting as a benchmark for future studies. Incidentally, in his preface, he acknowledged the help and encouragement of numerous people; his list reads like a 'Who's Who' of twentieth century ornithology!

We now come to two of the most important advances of the twentieth century. The first was the publication of *The Fauna and Flora of the Ilfracombe District of North Devon* (Palmer, 1946). The synopsis of this corner of Devon also included all the Lundy records known to the authors and, in particular, to the editor Mervyn Palmer who was Curator of the Ilfracombe Museum and Chairman of the Ilfracombe Field Club. It is primarily due to Palmer's enthusiasm and leadership that this remarkable compilation was achieved. Despite the fact that there was very little new information on Lundy, the book stood for many years as a handbook - a first reference to turn to for information on the island's natural history. Undoubtedly, it also helped to stimulate interest in the island. I first visited Ilfracombe Museum in 1948 and my interest in Lundy stemmed from that visit, the book which I purchased, and the tantalising views of the island from the mainland.

The second, even more important step was, of course, the formation of the Lundy Field Society in 1946 and the publication of the Annual Reports which commenced from 1947. From this point to the present day, a wealth of information on all aspects of Lundy's natural history has been published.

The second half of the twentieth century
The Annual Reports include many bird records and also the records and results of very many sessions of bird ringing. The published data on ringing (Taylor, 2004) lists almost 82,000 birds of 170 species, a remarkable effort. Many bird surveys have been carried out over the years, particularly of the island's very important seabird community. For example, in 1953, the Annual Report drew attention to the

increasing incidence of oiling of auks and gannets in the Bristol Channel, one of a series of reports which helped to ease the problem by new legislation.

Perhaps one of the more surprising aspects of studies of the Lundy fauna has been the papers published on the parasites found on the island. Between 1954 and 1988, a total of fifteen papers were published by six different authors covering ectoparasites (fleas, flat flies, lice and ticks) and endoparasites (roundworms, tapeworms and flukes) and are listed in Parsons, 1997. The amount of work involved was prodigious. For example, in the five years between 1952 and 1956 inclusive, 1252 ticks were collected from birds on the island.

One of the most encouraging aspects of recent years has been the study since 1993 of the ecology and conservation of Lundy cabbage and its associated fauna, culminating in papers in 1998 (Compton & Key) and 2000 (Key, Compton & Key). The regular presence of three prominent naturalists has also lead to many incidental records and their continuing work is leading to considerable advances in our knowledge of Lundy's ecology.

Some of the major studies have only taken place very recently. Although the lichens were reported in several papers between 1948 and 1972, it was not until the work of James, Allen & Hilton (1996 & 1997) that a comprehensive study was published. With the fungi, the comparable study of Hedger & George was published as recently as 2004 (Hedger & George, 2004). However, both studies are far from just species lists, providing large amounts of ecological data and setting extremely important baselines for the future.

Conservation

To warrant specific conservation measures for a site, one must first know what is present and which species are of particular importance, so the first requirement is a well-recorded site. By the 1970s, very considerable amounts of data had been collected and most of this had been published. The 1970s were a turning point in Britain, a time when most naturalists realised that specific measures were necessary to conserve our most important sites and species. The wheels of 'quangos' turn slowly and most of Lundy was eventually notified as a Site of Special Scientific Interest in 1987. In the meantime, the National Trust had purchased the island in 1969, thanks to the immense generosity of Jack Hayward, and Lundy had immediately received a good degree of protection from the Trust's own bye-laws. The interest being shown in the conservation of the island also encouraged Devon County Council to make several specific declarations from 1981, all these restraining future development. Thus, the island has various forms of statutory protection and should, in theory, be sacrosanct. However, we have all seen how governments are prepared to sacrifice areas of statutory protection on the altar of expediency and nobody in the conservation movement can afford to be complacent.

Of course, it is not only officialdom that can damage a site. One errant individual can do enormous damage. In the 1983 Annual Report, I commented on the presence of myxomatosis on the island (Parsons, 1984), assessing the arrival of the disease as probably being due to a deliberate introduction. Two years later, I was told who had introduced it - one cannot keep secrets on a small island. The results are difficult to assess, although one recent paper (Compton *et al.*, 2004) introduces a possible effect but the point is that the introduction should not have happened. Compare this with the actions to eliminate rats or to eradicate rhododendron which have been the subjects of careful research, consultation and monitoring. Everyone with an interest in the conservation of Lundy must be aware of the dangers of arbitrary introductions or unregulated extinctions.

THE FUTURE

Much of our information on the terrestrial ecology relates to the identification of species and an assessment of status where possible. Such data, including regular updates, will always be required in order to pursue conservation issues but, increasingly, studies will take place relating to the ecology of the island communities. Continuing to plot the origins and destinations of migrant birds through ringing; the assessment of the effects of climate change from distributional and phenological data; the rearing of invertebrates from substrates and hosts; the use of DNA to assess relationships; all these will be important in the future. It is essential, of course, that as much material as possible should be published.

A recent paper in the Annual Report, giving details of a survey of the microbial parasites of the brown rat on Lundy (Blasdell & Read, 2004), demonstrates several points. Firstly, many studies now involve experts using extremely sophisticated techniques. Secondly, the work is a part of a much larger study relating to reservoir hosts for specific viruses. Thirdly, the rats which were examined were available as a result of the eradication programme; the carcasses were not wasted.

In 1951, I joined the R.S.P.B. when it had about 5,000 members. Since then, the population of the U.K. has risen by about 20% and the membership of the R.S.P.B. by 20,000%. This reflects the enormous surge of interest in and concern for our environment and its species during the past half century, relating to improved education, increased leisure and greater disposable income.

I doubt whether anyone has worked out just how much the ancient murrelet was worth to Lundy but the income from transporting and feeding 5,000 birdwatchers must have been considerable. Less dramatically, if Lundy can keep its fauna and flora despite the increasing despoliation and decreasing wildlife elsewhere, then the income generated will help with the running costs of the island and the maintenance of the necessary conservation measures.

Chanter made the comment 'this remarkable Island, as a field for research ... is yet far from being worked out'. That is as true today as it was in the nineteenth century, and will still be true in the twenty-second.

REFERENCES

Blasdell, K. & Read, A. 2004. A survey of microbial parasites in Lundy brown rats (*Rattus norvegicus*). *Annual Report of the Lundy Field Society 2003*, 53, 86-98.
Brendell, M. 1976. Coleoptera of Lundy. *Annual Report of the Lundy Field Society 1975*, 26, 29-53.

Bristowe, W.S. 1929. The Spiders of Lundy Island. *Proceedings of the Zoological Society of London*, 2, 235-244.

Bull, S.A. & Parker, R.D. 1997. A Study of the Pygmy Shrew *(Sorex minutus)* on Lundy, 1996. *Annual Report of the Lundy Field Society 1996*, 47, 50-55.

Camden, W. 1607. *Britannia*. 6th edition. London: Newbery.

Chanter, J.R. 1887. *Lundy Island, A Monograph*. 2nd edition. London: Cassell, Petter & Galpin.

Compton, S.G. & Key, R.S. 1998. *Species Action Plan: Lundy Cabbage (Coincya wrightii) and its associated insects*. Peterborough: English Nature.

Compton, S.G., Key, R.S. & Key, R.J.D. 2004. Lundy cabbage population peaks - are they driven by rabbits and myxomatosis? *Annual Report of the Lundy Field Society 2003*, 53, 50-56.

Gardner, Keith 1968. Lundy - a Mesolithic Peninsula? *Annual Report of the Lundy Field Society 1967*, 18, 24-27.

Gosse, P.H. 1865. *Sea and Land*. London: James Nisbet.

Hedger, J.N. & George, J.D. 2004. Fungi on Lundy 2003.*Annual Report of the Lundy Field Society 2003*, 53, 62-85.

James, P.W., Allen, A. & Hilton, B. 1996. The lichen flora of Lundy: I: the species. *Annual Report of the Lundy Field Society 1995*, 46, 66-86.

James, P.W., Allen, A. & Hilton, B. 1997. The lichen flora of Lundy: II: the communities. *Annual Report of the Lundy Field Society 1996*, 47, 93-126.

Key, R.S., Compton, S.G. & Key, R.J.D. 2000. Conservation studies of the Lundy cabbage between 1994 and 2000. *Annual Report of the Lundy Field Society 1999*, 50, 49-59.

Langham, A.J. 1994. *The Island of Lundy*. Stroud: Alan Sutton.

Longstaff, G.B. 1907. *Lepidoptera and other insects observed in the parish of Mortehoe, N. Devon* (3rd edition). London: Mitchell Hughes & Clarke.

Palmer, M. (ed.) 1946. *The Fauna and Flora of the Ilfracombe District of North Devon*. Exeter: James Townsend & Sons.

Parsons, A.J. 1984. Notes on some mammals on Lundy. *Annual Report of the Lundy Field Society 1983*, 34, 40.

Parsons, A.J. 1997. Lundy's Non-marine Invertebrates. In R. Irving, J. Schofield and C. Webster (eds.) *Island Studies: Fifty Years of the Lundy Field Society*. Bideford: Lundy Field Society.

Perry, R. 1940. *Lundy: Isle of Puffins*. London: Lindsay Drummond.

Taylor, A.J. 2004. Bird ringing in 2003. *Annual Report of the Lundy Field Society 2003*, 53, 44-49.

Wollaston, T.V. 1845. Note on the Entomology of Lundy Island. *Zoologist*, 3, 897.

Wollaston, T.V. 1847. Further Notes on the Entomology of Lundy Island. *Zoologist*, 5, 1753.

Wright, F.R.E. 1933. Contribution to the flora of Lundy Island. *Journal of Botany*, November 1933, Supplement 1-11.

Wright, F.R.E. 1935. On the origin of Lundy flora, with some additions. *Journal of Botany*, April 1935, 90-95.

Zeuner, F.E. 1950. *Dating the Past: An Introduction to Geochronology*. 2nd edition. London: Methuen.

THE MACROFUNGI OF LUNDY

by

JOHN N. HEDGER,[1] J. DAVID GEORGE,[2] GARETH W. GRIFFITH[3] and LEWIS DEACON[4]

[1] School of Biosciences, University of Westminster, 115 New Cavendish Street,
London, W1M 8JS
[2] Natural History Museum, Cromwell Road, London, SW7 5BD
[3] Institute of Biological Sciences, University of Wales, Aberystwyth, SY23 3DD
[4] National Soil Resources Institute, School of Applied Sciences,
Cranfield University, Cranfield, MK43 0AL
[1] *Corresponding author, e-mail: hedgerj@wmin.ac.uk*

ABSTRACT

The chapter reviews the biodiversity and ecology of fungi on Lundy island, the chief focus being on the 'mushrooms and toadstools' or macrofungi, but some of the microfungi, especially plant pathogens, are included. The information used in the paper is derived from records published in the Annual Report of the Lundy Field Society from 1970 onwards, together with the results of brief surveys carried out by the authors between 2003-06. It is concluded that, although Lundy is, predictably, depauperate in species of fungi associated with woodland, the high diversity of fungi characteristic of unimproved grassland and heathland indicates that Lundy may be a site of national and even international importance. Suggestions are made for further work to confirm this status and for management strategies to maintain it.

Keywords: *Lundy, fungi, ecology, biodiversity*

INTRODUCTION

The History of Fungi on Lundy

The annual crop of edible macrofungi must have always been a welcome addition to the limited diet of the Lundy island community since its settlement by man. Some must have been put to additional uses, for example Bronze Age use as tinder for fire from some of the bracket fungi (Polyporaceae), although there is no archaeological record to substantiate this statement. No doubt both the Marisco and Heaven families enjoyed their mushroom feasts in the right season during their suzerainties, and, in more recent history, Diana Keast (personal communication 2005) remembers with pleasure dishes of 'field mushrooms' (*Agaricus campestris*) and 'wood blewitts' (*Lepista*) when she lived on the island. Some of the current inhabitants still continue to enjoy this annual bounty, especially the large 'parasol mushroom' (*Macrolepiota procera*) (Plate 1).

Plate 1: *Macrolepiota procera* (parasol mushroom) with the Old Light in the background. October 2003. *(Photo: John Hedger)*

More formal records of fungi on Lundy only began with reports of sightings, usually of the macrofungi, by interested visiting members of the Lundy Field Society and these have appeared at irregular intervals in the Annual Reports of the Lundy Field Society. The accumulated list for the period 1970-1995, abstracted from the Annual Reports, was summarised in Hedger & George (2004) and stood at some ninety-five species, to which they were able to make seventy-five additions from a week long survey in October 2003. Since then we have carried out further short surveys, in November 2004, April 2005 and January 2006 (Hedger, George, Griffith & Deacon unpublished data). In these surveys the field of search was extended to the microfungi on living plants, dead wood and plant litter, and 188 additional records were made, bringing the total to 358 species. It is planned to continue to publish these data in the Annual Reports.

The object of these studies was to start a more systematic inventory of the diversity of fungi on Lundy, and the habitats they occupy on the island, and to begin a database of Lundy fungi for entry in the British Mycological Society U.K. recording scheme. As with most mycological surveys, we have used the identification of fruit bodies of the fungi to establish the records, a practical approach, which can also be used to study the ecology of different species, but with some reservations, because the active mycelium remains hidden in the soil, wood or litter. Ecological surveying in this way is equivalent to using the flowers to map plant distribution, and has its predictable defects, especially in studies of the larger fungi. In some years fungi may not fruit, in others be abundant, which, combined with the shortness of study visits to the island, means that there is a high degree of serendipity to the process. In addition, the many microfungi are much more difficult to survey in this way, although determination of their fruiting structures with hand lens and microscope on particular plants or litter indicates their ecological preferences. However, despite these problems, even brief surveys can give useful information on the ecology of fungi on Lundy, especially their association with particular habitats and plants, and this is explored in more detail in this review.

The Lundy Climate and Fungi

There have been very few mycological studies of isolated British islands such as Lundy, one exception being the Hebrides from which Dennis (1986) recorded 2,905 species, although these have a much greater land area, and a different climate and soil to Lundy. Better parallels are, perhaps, the nearby Welsh islands of Skomer and Skokholm, but these are also little known mycologically. As with other aspects of island biology, there are intriguing contrasts to the mainland. The ameliorating influence of a maritime climate makes winter frost a comparative rarity, so that the macrofungi of autumn may continue to fruit for far longer on Lundy than the mainland, and on our visit in January 2006 we found quite a number of autumnal grassland species, long gone from the pastures of Devon. On the other hand, the strong winds on Lundy, the lack of shelter, and the thin granitic soils, may create very dry conditions in summer and early autumn, which reduce the fruiting of the fungi. Visits in September and October to search for macrofungi may be disappointing when compared to the mainland (Hedger & George, 2004). Our own experience indicates that early November is likely to be the best season, when soil temperatures remain warm, but moisture content has risen. However a wet summer would completely alter this picture.

The Roles of Fungi in the Terrestrial Ecosystem on Lundy

Fungi play a key role in the terrestrial food web on Lundy, as in all terrestrial ecosystems. One functional grouping, decomposer or saprotrophic fungi, are of great importance in the recycling of nutrients. They do so using enzymes which break down the components of plant litter and wood, eventually releasing CO_2, water and minerals. Their hyphae are in turn grazed by soil-, wood- and litter-inhabiting animals (detritivores), such as worms, thrips, mites and millipedes, which in turn form part of a food web which ends with the larger animals on Lundy, such as the pygmy shrew and many species of bird. On Lundy the most obvious decomposer species are many of the macrofungi, basidiomycetes and ascomycetes, both in the wooded areas and in different types of grassland and heath, but there are also many microfungi, some highly specific to litters of particular plant species, others much more widespread.

A second functional grouping is the many species of fungi which form mutualistic relationships with the roots of plants, termed mycorrhizas, and whose hyphae assist plants by uptake of key nutrients from the soil and litter, thus completing the recycling of minerals. On Lundy most of these 'helper' fungi are likely to be microfungi, forming single spores in the soil, and which infect the roots of herbaceous plants and grasses as Arbuscular Mycorrhizas (AM), although as yet no investigation has been made to prove their presence. Most belong to the family Engonaceae and a few, in the genus *Endogone*, form tiny, but visible, truffle-like fruit bodies, one of which was found in the 2003 survey in a rush clump (Hedger & George, 2004). A number of the trees on Lundy, including the sycamores *(Acer pseudoplatanus)* and ashes *(Fraxinus excelsior)* in Millcombe valley have similar AM partners to the grasses. However some of the other tree species on Lundy, for example all the species of oaks and pines, form associations called Sheathing or

Ectomycorrhizas with macrofungi (ascomycetes and basidiomycetes). These are sometimes specific to tree species, and are signalled by the presence of fruit bodies of the partner fungi under the tree in the autumn. These are produced by a mycelium which ramifies through the soil and litter, but which also envelopes part of the root system in a mycelial sheath, through which nutrient exchange takes place, thereby assisting the host tree.

A third functional grouping is the pathogenic fungi which invade living plants and trees on Lundy. The effects of these pathogens range from killing parts of the plant, to almost symptomless infections. Fungi which invade the plant and kill the tissue are often called 'necrotrophs' (lit. 'feeding off death') and their presence on plants is shown by brown areas of dead tissue on leaves and stems. The effects are often only local - for example, just a few brown spots on the leaves. On Lundy most of these necrotrophs are microfungi, some of which are specific to particular plant species, others have a wide host range. In contrast the fungal pathogens called 'biotrophs'(lit. 'feeding off life') infect the host with few or no symptoms. Infected plants may appear completely healthy, although there is sometimes some malformation of tissues. The production of spores by the fungus, for example in pustules on a leaf surface, may be the only sign that the plant is infected. These fungi have a very narrow host range, sometimes just one plant species. All are microfungi and on Lundy include the rusts (Uredinales), the smuts (Ustilaginales), the powdery mildews (Erysiphales) and the downy mildews (Peronosporales).

WOODLAND HABITATS FOR FUNGI ON LUNDY
The past land-use on Lundy, as with other small islands, has had a disproportionate influence on the vegetation and its associated fungi compared to the mainland. Clearance of such postglacial forest cover as existed on Lundy probably began in the Bronze Age and Hubbard (1971) considers that cutting of trees for construction and firewood by islanders and visiting ships, combined with the exposure, meant that the island was almost completely treeless as far back as the thirteenth century, although she cites evidence that scrub, such as gorse, willow and blackthorn, must have persisted. Large scale replanting of trees only began in the nineteenth century, many of them exotic species such as turkey oak, holm oak and Corsican pine, possibly accompanied by associated fungal species. The surviving tree cover is now restricted to the S.E. end of the island, especially the Millcombe Valley area (South Wood, North Wood, St John's Copse and Lower Millcombe), the small pockets of planted alders, sycamores, pines and oaks further up the east coast, including St Helen's Copse and Quarter Wall Copse, and the few scattered willows around the Quarry. As well as such 'true' woodland, the extensive gorse/blackthorn scrub behind Brambles Cottage and around the Flagstaff and the Ugly, as well as the rhododendron along the east coast path, also represent a type of woodland habitat (Dawkins 1974 in Hubbard 1997).

It has been estimated that about 80% of the macrofungi of the Netherlands are associated with trees (Arnolds & De Vries 1989 in Griffith *et al.*, 2004), and the U.K. figure is probably similar. Most of the decomposer and mycorrhizal fungi associated with the original forest cover on Lundy must have been lost, although Hedger & George (2004) have speculated that some of the wood decomposer species which they recorded fruiting on gorse and blackthorn on Lundy today, may represent persistence. Other species may have been introduced on the roots of the replanted trees or arrived as spores from the mainland. Of the 358 species of fungi at present recorded for Lundy only 65 were associated with trees, of which twelve were mycorrhizal species, the other 53 were wood and woodland litter decomposer species.

Wood Decomposer Fungi

The majority of the 53 species of decomposer fungi so far recorded in wooded areas were found fruiting on dead trees or fallen branches (wood decomposers), the rest were recorded from leaf litter or small twigs under the trees (litter decomposers). Commonest were the basidiomycete fungi popularly known as bracket fungi or polypores, but there were also species of other basidiomycetes, the agarics or gill fungi, and also of the sac or flask fungi, ascomycetes which also fruit on wood.

Primary attack on wood by fungi usually results in either bleaching (white rot decay) or the wood becomes brown and powdery (brown rot). The majority of wood decay on Lundy is by white rot fungi and the commonest species is undoubtedly the basidiomycete *Schizopora paradoxa*, (Plate 2) which can be found as a white crust with a surface of beautiful toothed pores, on rotting branches of nearly all tree and shrub species, even including rhododendron. Some other white rot basidiomycetes such as the 'zoned polypore', *Trametes versicolor*, also have a wide host range. In contrast other species have a narrow host range, for example, the aptly named 'blushing bracket', *Daedaleopsis confragosa*, found only on willows (*Salix* spp.), for example in the Quarries; the glistening white mushrooms of 'beech tuft' or 'porcelain fungus', *Oudemansiella mucida*, only on branches of the solitary beech *(Fagus sylvatica)* at Quarter Wall Copse; and the white crust-like *Lyomyces sambuci* (Plate 3) only on dead elder, *Sambucus niger*, especially in the Walled Garden at Millcombe. As already noted, Hedger & George (2004) speculated that the host preferences of many wood rot fungi on Lundy were different to the reports of 'normal' hosts in the literature, perhaps due to lack of the trees. One examples is *Phellinus tuberculosus*, (Plate 4) whose large hoof-shaped woody fruit bodies are normally found on *Prunus* and other rosaceous trees, but is often found fruiting on wood of dead gorse, *Ulex europaea*, on Lundy. Another is the 'beef steak mushroom' *Fistulina hepatica*, normally on oak, but found on sweet chestnut in the Millcombe Valley.

The ascomycete fungi involved in primary white rot attack on wood on Lundy include a number with hard black fruiting bodies - often called pyenomycetes. 'King Alfred's cakes', *Daldinea concentrica* (Plate 5), whose black fruit bodies are restricted to the ash trees *(Fraxinus excelsior)* in Millcombe valley, is a good example. Others include species of *Hypoxylon*, with groups of small hard rounded fruit bodies, for example *H. multiforme* on rhododendron wood, and species of *Xylaria*, with elongated black fruit bodies, such as the finger-shaped *X. polymorpha*, restricted to dead sycamore branches in Millcombe valley. Other ascomycete decay fungi appear on the underside of well-rotted branches and logs as small disc-shaped fleshy fruit bodies or

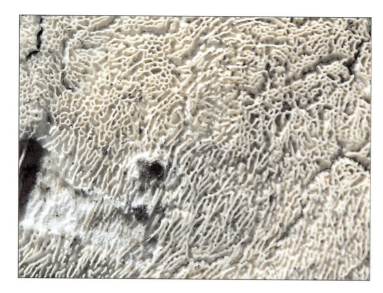

Plate 2: *Schizophora paradoxa*. Detail of fruit body on the underside of dead rhododendron log, east coast path. April 2005. ×2.5 *(Photo: David George)*

Plate 3: *Lyomyces sambuci* on dead elder in the Walled Garden, Millcombe Valley. April 2005. ×1 *(Photo: David George)*

Plate 4: *Phellinus tuberculosus* on dead gorse, behind Brambles. October 2003. ×0.5 *(Photo: John Hedger)*

Plate 5: *Daldinia concentrica* (King Alfred's cakes) on dead ash in Millcombe Valley. April 2005. ×1 *(Photo: David George)*

Plate 6: *Mollisia cinerea* on dead willow in the Quarries. April 2005. ×1 *(Photo: David George)*

Plate 7: *Hypholoma fasciculare* (sulphur tuft) on dead rhododendron near The Ugly. October 2003. ×0.75 *(Photo: John Hedger)*

apothecia (discomycetes). Unlike the pyenomycetes they do not seem to be usually primary decomposers, but colonise wood in an advanced state of decay. The commonest are grey apothecia produced by *Mollisia* species (Plate 6). These may occur alongside the attractive bright yellow apothecia of *Bisporella citrina*.

Others of these minute discomycetes occur on specific hosts, for example troops of the beautiful, but minute, white discs of *Hyaloscypha stevensonii* are only found on undersides of well-decayed fallen branches of Corsican pines, *Pinus nigra* var. *maritima*, in the Millcombe Valley. Alongside these discomycetes on the well-rotted pine wood can be found another group of organisms, the myxomycetes, or slime moulds, which although protistans, and not true fungi, produce minute fungal-like fruiting bodies. Their plasmodia feed on the previous fungal and bacterial colonisers in the wood prior to fruiting. Species found to date include the pink *Arcyria denudata*, the intriguingly basket-shaped *Comatricha nigra* and the yellow pin-headed *Trichia varia*. On the drier upper surface of the pine logs, the confusingly similar, yellow, button-shaped, minute gelatinous fruit bodies of *Dacrymyces stillatus* may be found. This is actually a basidiomycete fungus. The fruit bodies are very tolerant of drying and the species is also a successful coloniser of the tops of fence posts and gates all over the island. It is closely related to the 'jelly fungus', *Tremella mesenterica*, whose orange gelatinous fruit bodies occur on dead wood of gorse, and to the 'ear fungus', *Auricularia auricula-judae* which colonises dead elder and is common in the Walled Garden in Millcombe Valley. Interestingly these three gelatinous species are amongst the earliest records of fungi from Lundy (Walker & Langham, 1971), probably because of the durability of their fruit bodies and lack of seasonality.

There are fewer fungi on Lundy which cause brown rot decay of wood, the most common being the yellow to orange-brown, crust-like fruit bodies of the 'polypore' *Coniophora puteana* and the related species, *C. arida*, which are very abundant on the many dead treated stumps of rhododendron by the side of the coastal path in the east side clearance areas, and which are slowly reducing these stumps to brown powder. Normally these species are recorded from conifer wood. Interestingly, many rhododendron stumps are also colonised by white rot fungi. These include the 'sulphur tuft' agaric, *(Hypholoma fasciculare)* (Plate 7) which forms clumps of sulphur yellow toadstools. Where the species occur together, contrasting brown and white rot can be seen in the same stump. In passing it should be noted that the bonfires in this area have stimulated the autumn fruiting of troops of a small brown-capped agaric, *Pholiota carbonaria*, which is restricted to burnt ground.

Mycorrhizal Fungi and Woodland Litter Decomposer Fungi
Under the trees, the fungi that are found fruiting on the soil or litter are either mycorrhizal, associated with the roots of the trees, or are decomposer fungi, breaking down the leaf and twig litter. The mycorrhizal species recorded to date are large and obvious late summer/autumn macrofungi, including agarics and boletes in the genera *Lactarius*, *Russula*, *Cortinarius*, and *Boletus* and the 'earth ball' (*Scleroderma citrinum*). They represent some of the earliest observations of fungi on Lundy in the early 1970s (Walker & Langham, 1971; Walker, 1972). Examples found in our 2003

survey (Hedger & George, 2004) included the dull yellow agaric *Russula ochroleuca* under the holm oaks in St John's Copse, and the brighter purple *R. krombholtzii* (=R.atropurpurea) under the beech in Quarter Wall Copse growing alongside the yellow-pored bolete *Xerocomus (=Boletus) spadiceus*. Our own records and past records combine to give a total of some twelve species of mycorrhizal macrofungi, a strikingly small total compared to a mainland woodland. Perhaps Lundy is truly depauperate in mycorrhizal fungi, but more must be present but not yet recorded. For example *Amanita muscaria*, the 'fly agaric', a large fungus with bright red cap with white spots, would be expected in association with the planted species of pine *(Pinus radiata, P. nigra* var. *maritima* and *P. sylvestris)*, as would the large bolete *Suillus luteus*, 'slippery jack', and yet neither have been recorded. This may be true of other almost universally common mycorrhizal species in the U.K., such as the orange brown agaric *Laccaria laccata*, 'the deceiver' and the related purple *L. amethystina*, 'the purple deceiver', both yet to be found on Lundy.

The litter under the trees is broken down by decomposer species of fungi. As their mycelium spreads through the litter it is often bleached by their action. As with the mycorrhizal fungi, some are associated with particular tree species, and may also be at least in part mycorrhizal. The best example is the small brown agaric *Alnicola suavis*, appropriately found growing on the leaf litter under the alders in St Helen's Copse, but also with the willows at the Quarry (Hedger & George, 2004). Other species of decomposers are more widespread and include a number of species of delicate, often small, white-gilled agarics in the genus *Mycena*, (Plate 8) and medium sized species of *Collybia*, also with white gills, but with flat white to brown caps and tough bendable stems, such as the dry-capped *C. dryophila*, also found decomposing bracken litter all over Lundy, and the 'buttery agaric', *C. butyracea*, a greasy-capped, late autumn species preferring acid litter, and found fruiting under the pines and gorse in Millcombe Valley. The larger woodland litter decomposer agarics include the 'shaggy parasol mushroom', *Macrolepiota rhacodes*, found in November 2003 fruiting in the nettles by the Castle (Hedger & George, 2004), but probably also present in the Millcombe Valley. This is similar to the 'parasol mushroom', *M. procera*, widespread in autumn in the grassland around the Old Light, but differs in having a scaly stem and in slow reddening of the flesh of the stem when bruised or cut.

Plate 8: *Mycena filopes* under willow in the Quarries. October 2003. ×2 *(Photo: John Hedger)*

GRASSLAND AND HEATH HABITATS FOR FUNGI ON LUNDY

Grassland and Heath vegetation covers most of Lundy and is the major terrestrial habitat. It is not, however, uniform. Hubbard (1997) divided the non-woody vegetation on Lundy into two categories: the Sidelands and the Plateau, and the same approach is followed here.

The Sidelands

Of the Sidelands, the east coast has the most species-diverse plant communities, especially on the shale slopes of the southern end, where the Lundy cabbage, *Coincya wrightii* has its stronghold. Moving further north along the east coast path into the granite geology, the dominance of rhododendron and then bracken reduces the diversity of plants and associated fungi, although there are a number of very interesting fungi in these localities associated with the plant cover, for example host-specific fungi on the rhododendrons, such as 'purple leaf spot' *Cercoseptoria handelii* and the aptly named 'bud blast', *Pycnostysanus azaleae*, which can be detected by its tiny pin-shaped spore structures on dead blackened flower buds. In contrast, the west and north coast Sidelands are much more exposed and botanically poorer, being dominated by just a few species, especially sea fescue *(Festuca rubra)* and sea pink *(Armeria maritima)*, so it is also likely to have fewer species of fungi.

Most of the species of macrofungi found on the granitic areas of the East Sidelands are also found on the Plateau, for example those associated with bracken, which dominates sideland areas such as Brazen Ward. These are mostly decomposer agarics growing as mycelia in the deep bracken litter, such as the bright orange *Hygrophoropsis aurantiaca*, 'the false chanterelle', (Plate 9) and grey-white species of *Collybia* and *Clitocybe*, the latter genus including the large, stoutly-stemmed, *C. nebularis*, 'the clouded clitocybe', which forms fruit body rings 2-3 m in diameter in the autumn in the bracken. In places where wet peaty soil occurs, especially in eroded exposures alongside the coast path around Gannets Bay, the tiny, but beautiful, pale-yellow agaric *Omphalina (=Gerronema) ericetorum* (Plate 10) fruits in small groups and was found to be abundant in our survey in April 2005. This fungus is associated with an alga which grows on the peat surfaces (=the lichen genus *Coriscium*). Areas of grass inside the bracken stands had some of the grassland species also found on the plateau, but the beautiful yellow coral-like fruit bodies of *Clavulinopsis corniculata* were more abundant on the East Sidelands than on the Plateau in November 2004. A very unusual dwarf form of this fungus (Plate 11) was also found at the same time on the West Sidelands in *Festuca rubra* turf near the North Light. Whether this is more widespread along the West Sidelands, or indeed if there is a distinct decomposer community in the *Festuca rubra/ Armeria maritima* turf is yet to be determined. The decomposers like *Clitocybe* and *Collybia* found in the East Sidelands were not recorded in our brief surveys along the west coast: the small yellow/brown agarics in the genus *Galerina*, were the only common macrofungi, especially *G. hypnorum*, (Plate 12) fruiting amongst polytrichum moss.

The majority of past records of plant pathogens are also from the East Sidelands (Hedger & George, 2004), reflecting the greater plant diversity. The rusts, (Uredinales)

(Above) **Plate 9**: *Hygrophoropsis aurantiaca*
(false chanterelle) under bracken, near Brazen
Ward. October 2003. ×1 *(Photo: John Hedger)*

(Right) **Plate 10**: *Omphalina (Gerronema)*
ericetorum on wet peat. Coast path, Gannets Bay.
April 2005. ×2 *(Photo: David George)*

Plate 11: *Clavulinopsis*
corniculata, dwarf form.
In *Festuca rubra* turf
near the North Light.
November 2004. ×1
(Photo: Gareth Griffith)

Plate 12: *Galerina*
hypnorum. In moss on
the edge of the path to
the Battery. November
2004. ×3 *(Photo: Gareth*
Griffith)

are often host-specific, but may form different spore types on two alternate hosts. The most productive time of year to survey these fungi is in the late winter, spring and early summer and our own visits in April 2005 and January 2006 have greatly extended the records of these fungi on Lundy, but many have yet to be found. One of the most obvious is *Phragmidium violaceum* which is common on bramble, *Rubus fruticosus* agg., as obvious purple spots on the upper leaf surface, below which yellow sporing structures (aecidia) are formed in the spring, followed by purple-black spore patches (telutosori) in summer and autumn. *Puccinia smyrnii* (Plate 13) can be easily found on leaves and stems of all the plants of the attractive yellow-flowered umbellifer, alexanders *(Smyrnium olusatrum)* in the lower Millcombe Valley. It forms pustules on which are yellow (aecidia) and later brown (uredosori) associated with deformed, swollen growth. Aecidia of *Uromyces dactylidis* (Plate 14) are of similar appearance on leaves of celandine (*Ranunculus ficariae*) also in Millcombe, whilst *Uromyces muscari* shows up as darker telutosori on the bluebell *(Scilla non-scripta)* leaves. Stinging nettles *(Urtica dioica)* infected by *Puccinia caricina* also show up in spring because of deformed growth of stems and leaves bearing bright orange aecidia. The spores from these plants infect an alternate host, species of sedge, *Carex,* in nearby grassland, on which other spore types are formed. Walking along the east coast path the bright orange aecidia of *Puccinia violae* are obvious on leaves of the dog violet *(Viola canina)*, accompanied by *Puccinia umbilici*, signalled by bright red spots on the leaf upper surface, with black telutosori below, on leaves of the wall pennywort *(Umbilicus rupestris)*. Less obvious, and infrequently recorded, are the rusts on ferns, belonging to the genus *Milesina,* which form white uredospores in spring on the old over-wintering fronds. We recorded three species of these rusts in our visit in January 2006: *M. dieteliana* was found on the clumps of common polypody *(Polypodium vulgare)* growing in the barn wall of Lundy Farm; on the clumps of ferns on the faces inside the Quarries we found *M. kriegeriana* on the broad buckler fern *(Dryopteris dilatata)*; *M. scolopendrii* on leaves of the harts tongue fern *(Phyllitis scolopendrii)*.

The smuts (Ustilaginales) are a group of plant pathogens related to the rusts, but which have yet to be recorded from Lundy. One example is *Ustilago violacea*, the 'anther smut'. On Lundy this would be expected on the sea campion *(Silene maritima)* and the red campion *(S. dioica)*, where infected plants are only revealed by the replacement of pollen by purple brown masses of spores in the centre of the flower - in addition female flowers are converted to male flowers. In April 2005, in spite of intensive searching of *S. dioica* plants in the Millcombe Valley and the east coast path, and a binocular search of the isolated flowering clumps of *S. maritima* on goat inaccessible ledges along the west and east coast, no infected plants were found. This is surprising, since *U. violacea* is very common on the two campions in very similar habitats on Skomer Island (Hedger, unpublished data). It is possible that this species, as well as the squill 'anther smut' *(Ustilago vaillentii)* on spring squill *(Scilla verna)*, also common on Skomer, will be found on Rat Island, which is the only part of Lundy without significant grazing (Hubbard, 1997), and which would profit from a survey in April/May.

Plate 13:
Puccinea smyrnii, Aecidia on *Smyrnium olusatrum* (alexanders) in Lower Millcombe Valley. January 2006. ×4 *(Photo: David George)*

Plate 14:
Uromyces dactylidis, Aecidia on *Ranunculus ficaria* (celandine). Upper Millcombe Valley. April 2005. ×3 *(Photo: David George)*

Plate 15:
Possible *Mycosphaerella* sp. On *Coincya wrightii* (Lundy cabbage). Beach Road. April 2005. ×1.25 *(Photo: David George)*

However, during our search for *U. violacea* along the shale cliffs below and above the Beach Road in April 2005 we did find a number of other interesting pathogens on the cliff plants. On the lower cliff, in the splash zone, leaves of scurvy grass *(Cochlearea officinalis)* were heavily incrusted with white spores of downy mildew, *Albugo candida*. Higher up, records included dark patches on leaves of foxglove, *(Digitalis purpurea)*, caused by *Ascochyta molleriana*, and yellow patches on leaves of cuckoo pint *(Arum maculatum)* caused by *Ramularia ari*, both very common fungi on these hosts on Lundy. Of particular interest were pathogens on plants considered by Hubbard (1997) to be Lundy rarities: the *Phomopsis* state of *Diaporthe arctii* was found bleaching patches on leaves of the locally abundant balm-leaved figwort *(Scrophularia scorodonia)*, but the most exciting find was a species of *Mycosphaerella*, causing brown lesions (Plate 15) on leaves of the Lundy cabbage, *Coincya wrightii*. It is as yet undetermined, but is possibly *M. brassicola*, or an undescribed, host-specific species, and therefore a Lundy endemic. More material needs to be examined before this question can be answered.

The Plateau
On the Plateau, the very extensive areas of grassland, and heath, maintained by the intensive grazing and exposure on Lundy, represent a very important habitat for fungi, indeed one which is of significant conservation interest for the whole of the U.K.

Following Hubbard (1997) we have distinguished four main types of fungal habitat on the plateau: 1) short-cropped turf, comprising most of Middle Park, much of Ackland's Moor, the Airfield, the South West Field and Castle Hill; 2) taller and matted grassland, with *Molinia* and bracken, especially the area between Quarter Wall and Halfway Wall, with a subset of much wetter areas with *Sphagnum*, *Molinia* and *Juncus* around Pondsbury, in the Punchbowl Valley and in the shallow valley running down the east side of the island beside Quarter Wall to the Quarries; 3) the walled off areas of improved grassland around the farm; 4) the *Calluna vulgaris*-dominated areas north of Threequarter Wall. Of these habitats, the enclosed fields around the farm are most species-poor in both plants and fungi. In contrast, the most species-rich areas for macrofungi are the short-cropped turf, followed by the taller grassland and the *Calluna*. Most of the fungi recorded are decomposers, growing on plant litter or herbivore dung, but we did find one woodland mycorrhizal species, the 'cob web agaric' *Cortinarius anomalus*, associated with the clumps of willow and gorse by the Threequarter Wall gate on the east side of the island.

1. Short Turf Grassland
The decomposer macrofungi of the short cropped grassland are the most obvious to the visitor to the island in autumn, especially since they are abundant in the area around Lundy Old Light. However the nearby Airfield is another 'hotspot', as is the west side of Middle Park. They include large agarics such as the 'parasol mushroom', *Macrolepiota procera*, the 'field mushroom', *Agaricus campestris*, the 'horse mushroom', *Agaricus arvensis* and the purple coloured 'wood blewitt', *Lepista nuda*. In the short turf it is easy to see that many of these decomposer fungi grow through

the turf as 'fairy rings', obvious by the ring of fruit bodies formed at the edge of the growing mycelium, or as rings of darker green grass. The rings of *Lepista nuda* (Plate 16) are particularly abundant on the Airfield, but alongside can also be found rings formed by smaller agarics, the white, tough-stemmed, but edible, 'fairy ring champignon' *(Marasmius oreades)*, and the confusingly similar, but stouter-stemmed, 'false fairy ring champignon', *Clitocybe dealbata*, a poisonous species.

In addition there are a number of Gasteromycete ('puffball') species which grow in the short turf and these also sometimes form clear rings. They include *Vascellum depressum*, (Plate 17) a white, later brown, short-stalked puffball, 3-6cm diameter; *Bovista nigrescens*, a white, later black and leathery, stalk-less puffball, 2-5 cm diameter; and *B. plumbea*, with fruit bodies similar to *B. nigrescens*, but smaller, 1-3 cm diameter and turning from white to lead-grey as they mature. There are 'hot spots' for the autumn fruiting of these species - one in the very short turf on the left-hand side of the track just north of the gate through Threequarter Wall, another in the short turf by the Mangonel Battery at Threequarter Wall. However in winter the leathery fruit bodies detach from the soil and are blown all over the island, even ending up in the flotsam and jetsam on the beaches. In passing it should be noted that there are also two much larger grassland 'puffballs' on Lundy - the 'giant puffball' *Langermannia gigantea* was recorded by Walker & Langham (1971) in the S.W. Field, and we have found the empty stalked fruit bodies of *Calvatia utriformis* in the Graveyard.

Most of the decomposer fungi of the short turf grassland do not often form easily recognisable rings. These include *Hygrocybe*, a genus of sometimes strikingly brightly coloured agarics, popularly known as 'wax caps'. In our initial survey in October 2003 (Hedger & George, 2004) we found no *Hygrocybe* species, due to drought conditions that year, although 15 species had previously been recorded since 1970 by visiting members of the LFS (Hedger & George, 2004). In November 2004 we found 20 species, of which nine were new records for Lundy, bringing the total for Lundy to 24, making it the most species- diverse genus of macrofungus on the island. This total, which can be compared to the U.K. total of around 40 species (Griffith *et al.*, 2004), is of great interest, since in recent years the diversity of *Hygrocybe* species in grassland has been used as an indicator of habitat quality, in particular of low soil nutrient status, and lack of nitrogen enrichment from artificial fertilisers or pollution. Boertmann (1995) considers that an overall site species total for *Hygrocybe* of 17-32, with 11-20 being recorded in a single visit, is indicative of national conservation importance in Northern Europe, a figure we reached in November 2003. In addition Boertmann considers some *Hygrocybe* species to be more fastidious indicators than others. In our own survey in November 2004 we found that *H. virginea*, a white-capped species, which is quite tolerant of high soil nitrogen, (Griffith *et al.*, 2004), was the only species fruiting on the improved grassland of the farm fields. In contrast, in a survey of the short turf on the Airfield, and in Middle Park near the Mangonel battery, we found fifteen other *Hygrocybe* species in a search area of 250m^2. Large species included *H. coccinea* (scarlet), *H. punicea* (dark red), (Plate 18), *H. splendidissima*, (Plate 19), (scarlet and yellow)

Plate 16: Ring of *Lepista nuda* (wood blewitt) near the Airfield. November 2004. *(Photo: Gareth Griffith)*

Plate 17: *Vascellum depressum*. Old fruit body. The Airfield. January 2006. ×1 *(Photo: David George)*

Plate 18: *Hygrocybe punicea*. The Airfield. November 2004. ×1 *(Photo: Gareth Griffith)*

Plate 19: *Hygrocybe splendidissima*. The Airfield. November 2004. ×1 *(Photo: Gareth Griffith)*

Plate 20: *Hygrocybe pratensis*. The Airfield. November 2004. ×1.5 *(Photo: Gareth Griffith)*

Plate 21: *Hygrocybe conica*. The Airfield. November 2004. ×1.5 *(Photo: Gareth Griffith)*

and *H. pratensis* (Plate 20), (honey coloured). Smaller species included the bright green, later lilac or yellow, *H. psittacina*, *H. conica*, (Plate 21), orange-red, blackening with age, and the small white-cream species *H. russocoriacea*, which smells of leather. However, absence of improvement of the grassland is not the only factor affecting *Hygrocybe* species distribution on Lundy. In the much more acidic tall grassland north of Quarter Wall and around Pondsbury and just north of Threequarter Wall we found only *H. laeta*, a pinkish-brown medium-sized species, also found in the short turf areas, but presumably more tolerant of low soil pH. An even more interesting result in November 2004 was the finding of a medium-sized grey species of *Hygrocybe* fruiting abundantly on the thin peaty soil under the *Calluna* in the immediate vicinity of the Bronze Age fort at the North End of Lundy. This agaric was never seen in any other location on Lundy and is similar to *H. Radiata*, (Plate 22), a northern European species and one of the rarest *Hygrocybe* species in the U.K. (ca. 30 U.K. records in total). A dried specimen of this fungus has been deposited in the mycological herbarium at the Royal Botanic Gardens, Kew and will soon be subjected to critical determination by a specialist.

Plate 22: *Hygrocybe c.f. radiata*. Young fruit bodies. North End. November 2004. ×1.5 *(Photo: Gareth Griffith)*

Plate 23: *Entoloma (lampropus* group). The Airfield. November 2004. ×2 *(Photo: John Hedger)*

There are a number of other fungi which fruit alongside the *Hygrocybe* species in the short turf grassland on Lundy, especially on the Airfield. Significantly some of these have also in recent years been used, with the *Hygrocybe* species, as additional indicators of the conservation value of the site (Griffith *et al.*, 2004). The agarics include species in the genus *Entoloma*, medium or small species mostly characterised by grey-brown caps and pink gills, and to date nine species have been recorded, including *E. lampropus*, a beautiful agaric with a steel-blue cap and stem (Plate 23), and *E. staurosporus*, with a conical brown cap. Members of this genus are difficult to identify and it is very likely that more species remain to be discovered on Lundy. In addition, in November 2004, we found growing alongside these agarics other 'indicator' fungi belonging to the Geoglossaceae, a family in the phylum Ascomycota or 'sac' fungi. These fungi form small tongue-shaped fruit bodies. Two were black species of *Geoglossum*, *G. fallax* and *G. glutinosum*, the other the much rarer green-coloured *Microglossum olivaceum*. The presence of these fungi, together with *Hygrocybe* species, and other grassland fungi, has been used by Rotheroe (1996) and others to indicate the conservation value of a grassland site.

2. Tall Grassland

Rough grazing of this kind dominates the centre of the area between Quarter Wall and Halfway Wall. Compared to the short turf areas it has few species of macrofungi. A few of the short turf grassland species of macromycetes are found, but usually only in patches of drier ground that has been grazed. The most common fungi growing between the *Molinia* and *Juncus* clumps are species of the agaric genera *Galerina* and *Agrocybe*, small agarics with a reddish-brown conical cap and gills, for example *G. tibiicystis*, *G. vittiformis* and *A. paludosa*, whilst the appropriately named *G. sphagnorum* was found in the wetter areas around Pondsbury, fruiting alongside two larger, long-stemmed, dark-gilled and brown-capped species of *Hypholoma*, *H. subericaeum* and *H. elongatum*. A most interesting feature of this habitat is the humid microclimate afforded by the dense clumps of *Juncus* and, to a lesser extent, *Molinia*. In October 2003, in spite of the very dry conditions, many records were made of decomposers, mostly microfungi, growing on the litter in the centre of clumps (Hedger & George, 2004), and this has been extended on each visit since, even in winter - six more records were obtained from *Juncus* clumps in January 2006. On all occasions the most common fruit bodies found were the minute (1-2 mm diameter) pure white-stalked cups of *Dasycyphus apalus*, very common in the large *Juncus* clumps around Pondsbury, growing alongside the equally minute sessile grey cups of *Mollisia juncina*. The large *Carex paniculata* clumps along the streams in Gannets Combe, and in the St John's valley, are an interesting parallel habitat, but a brief examination in January 2006 showed that they contain different host related species of decomposer microfungi, for example *Mollisia caricina*. The same principle of host specificity applies to much of the plant litter on Lundy which is relatively slow to decay - other examples are the small elongated black fruit bodies of *Rhopographus filicinus* on dead petioles in bracken clumps, and the minute brown fruit bodies of *Leptopeltis nebulosa* on the dead fronds of the royal fern, *Osmunda regalis* growing on the walls of the Quarries.

One of the other features of the tall grassland is that the higher humidity favours the development of fungal fruit bodies on the herbivore dung, which tends to dry out in the short turf areas. Indeed the habitual presence of some of the Lundy ponies in a very rushy area near to the Quarter Wall gate makes this a productive area to search for coprophilous fungi. The very large quantities of herbivore dung deposited everywhere on Lundy are decomposed by specialised coprophilous fungi, many of which have spores which require passage through the gut of a herbivore before they will germinate. The species list of coprophilous fungi for Lundy remains small only because of lack of investigation (Hedger & George, 2004). Early colonisers are microfungi, such as the aptly named 'hat thrower' (*Pilobolus* spp.) which forms glistening mats on relatively fresh horse apples in the early morning and discharges its minute black sporangia onto surrounding grass. Older more decomposed horse dung is often covered with the minute orange disc-shaped apothecia of the ascomycete *Coprobia granulata*, and also produces large agarics such as the handsome egg-shaped capped *Anellaria semiovata*, and the pure white 'dung inkcap', *Coprinus niveus*. A related coprophilous agaric, *Paneolina foenisecii*, a small greyish-brown species with a rounded cap with a distinctly lighter edge, is probably the most common fungus on Lundy at all times of the year, and also occurs in large numbers in the short turf grassland, growing alongside the smaller pointed capped 'liberty cap' *(Psilocybe semilanceata)*. The 'field mushroom', *Agaricus campestris*, which, as already noted, is very abundant on Lundy, utilises well rotted dung, particularly horse dung, that has become incorporated into the soil. However its fruiting is restricted to the short turf area, and, in spite of the abundance of horse apples in the rushy pastures around Quarter Wall gate, we have never recorded it there.

3. Calluna Heath

The short heath of *Calluna vulgaris* and associated plants, which grows on the thin acid peat of the Plateau of the North End is very different to the rest of the island and remoteness has meant that past records, and our own surveys, have only involved brief visits. Some of the agarics recorded here are found elsewhere on Lundy growing on acidic litter under bracken, *Calluna*, or the *Molinia* areas, e.g. decomposer agarics such as the orange 'false chanterelle', *Hygrophoropsis aurantiacus*, and the cream-capped *Collybia dryophila*. However other fungi found here do not occur elsewhere on Lundy. *Collybia obscura*, a small purplish decomposer agaric with a tough flesh and unpleasant smell was only found at the North End in October 2003 (Hedger & George, 2004) and again in November 2004. Likewise, as noted earlier, the rare *Hygrocybe* cf. *radiata* was only recorded around the North End in November 2004. During the same visit, a spectacular myxomycete, *Leocarpus fragilis*, (Plate 24), was abundant, covering patches of *Calluna* with large bright yellow gelatinous plasmodia, later maturing to a mass of glistening dull red minute egg-shaped sporangia. Ing (1999) considers *L. fragilis* to be characteristic of the acidic litter of conifers and gorse, where the plasmodium migrates upwards to fruit on trunks well above the soil. No mention is made by Ing of *Calluna*, but the organism is obviously very successful in this habitat on Lundy.

DISCUSSION

It could be asked whether a study of fungi on Lundy has any scientific value. Any of the fungi recorded would be likely to be also present on the mainland. However, the account presented in this chapter highlights a number of counter arguments. Firstly the national conservation status of Lundy means that a complete inventory of the organisms present is highly desirable, and although the lichens, part fungus, part alga, have been previously well

Plate 24: *Leocarpus fragilis* (Myxomycete) fruiting on *Calluna*. North End. November 2004. ×2.5 *(Photo: Gareth Griffith)*

documented (Noon & Hawksworth, 1973; James *et al.*, 1996; 1997) the aim must be to raise the knowledge of Lundy fungi to the same status. As noted in the introduction to this chapter, the evanescent nature of fruit bodies of fungi makes this task difficult, but the present total of 358 species compiled from our own visits, and from previous records of LFS members, means that a good beginning has been made. Secondly, our studies have begun to reveal that there are some unique features to the ecology of fungi on Lundy, for example the unusual hosts for some of the wood-rotting fungi and the curious records of fungi from the North End. Perhaps there are 'island' effects on the fungi.

Thirdly, even if Lundy were an area of land on the mainland, the high diversity of the grassland fungi noted in this chapter would make a strong case for SSSI status. In the last 60 years there has been a loss of over 90% of species-rich grazing land and hay meadows in southern England, due to agricultural improvements, such as ploughing and reseeding, and application of fertilisers (Griffith *et al.*, 2002; 2004). Since the early '90s, surveys have shown that these remnants of unimproved grassland and heathland in the U.K. also support many rare species of macrofungi, so much so that two SSSIs have recently been designated to conserve these fungi (Rotheroe *et al.*, 1996; Rotheroe 2001). Many of these sites are tiny, including graveyards and lawns, but are chiefly characterised by low soil nutrient status and absence, in particular, of application of nitrogen. On Lundy a large proportion of the island is managed in ways which are favourable to these communities of fungi, with no fertiliser input and heavy grazing, the exception being the enclosed improved grasslands around the farm. The island should therefore be recognised as an important U.K. site for 'grassland' fungi. It is safe to predict that even more grassland species will be recorded in the future. The positive message is that the current management of Lundy is directed to conservation of biodiversity, and, if this includes the continuation of grazing of most of the island without fertiliser, whether organic or inorganic, these fungi are safe.

Ironically conservation activities on Lundy may also lead to the loss of some of the species of fungi! The current plans for eradication of the east coast rhododendron thickets may result in the loss of the dozen or so species of fungi presently recorded for Lundy from rhododendron, such as 'rhododendron bud blast', *Pycnostysanus rhododendri*, (Hedger & George, 2004). However, any loss is likely to be offset by an increase in plant biodiversity on the east side of Lundy, and perhaps re-introduction of native species of tree along the east side of the island, along with their associated fungi.

ACKNOWLEDGEMENTS

The authors wish to thank the Lundy Field Society and the British Mycological Society for grants which made the surveys possible and Graham Coleman, Jenny George and Brenda McHardy for their help with the field work. We also thank the School of Biosciences, University of Westminster for logistical and financial support, especially Graham Coleman for the loan of microscopes and other equipment.

REFERENCES

Boertmann, D. 1995. *The genus Hygrocybe*. Fungi of Northern Europe Vol. 1. Copenhagen: Danish Mycological Society.

Dennis, R.W.G. 1986. *Fungi of the Hebrides*. Kew: Royal Botanic Gardens.

Griffith, G.W., Easton, G.L. & Jones, A.W. 2002. Ecology and Diversity of Waxcap (*Hygrocybe* spp.) Fungi. *Botanical Journal of Scotland*, 54, 7-22.

Griffith, G.W., Bratton, J.H. & Easton, G.L. 2004. Charismatic Megafungi, the Conservation of Waxcap Grasslands. *British Wildlife*, 15(3), 31-43.

Hedger, J.N. & George, J.D. 2004. Fungi on Lundy 2003. *Annual Report of the Lundy Field Society 2003*, 53, 62-85.

Hubbard, E.M. 1971. A Survey of Trees on Lundy. *Annual Report of the Lundy Field Society 1970*, 21, 14-19.

Hubbard, E.M. 1997. Botanical Studies. In R.A. Irving, A.J. Schofield & C.J. Webster, *Island Studies, Fifty Years of the Lundy Field Society,* 141-148. Bideford: Lundy Field Society.

Ing, B. 1999. *The Myxomycetes of Great Britain and Ireland*. Slough: Richmond Publishing Co.

James, P.W., Allen, A. & Hilton, B. 1996. The Lichen Flora of Lundy. I. The Species. *Annual Report of the Lundy Field Society 1995*, 46, 66-86.

James, P.W., Allen, A. & Hilton, B. 1997. The Lichen Flora of Lundy. II. The Communities. *Annual Report of the Lundy Field Society 1996*, 47, 93-126.

Noon, R.A. & Hawksworth, D.L. 1973. The Lichen Flora of Lundy. *Annual Report of the Lundy Field Society 1972*, 23, 52-58.

Rotheroe, M., Newton, A., Evans, S.E., Feehan, J. 1996. Wax Cap Grassland Survey. *The Mycologist*, 10, 23-25.

Rotheroe, M. 2001 A Preliminary Survey of Wax Cap Species in South Wales. In: D. Moore, M.M. Nauta, S.E. Evans, M. Rotheroe (eds), *Fungal Conservation: Issues and Solutions*, 120-135. Cambridge: Cambridge University Press.

Walker, A.J.B. & Langham, M.S. 1971. Some Tentative Identifications of Lundy Fungi. *Annual Report of the Lundy Field Society 1970*, 21, 34-35.

Walker, A.J.B. 1972. Additional Fungi on Lundy May 1971. *Annual Report of the Lundy Field Society 1971*, 22, 43.

LUNDY CABBAGE: PAST, PRESENT, FUTURE

by

Stephen G. Compton[1], Jenny C. Craven[1], Roger S. Key[2]
and Rosemary J.D. Key[2]

[1] Faculty of Biological Sciences, University of Leeds, Leeds, LS2 9JT
[2] Natural England, Northminster House, Peterborough, PE1 1UA
Corresponding author, e-mail: S.G.A.Compton@leeds.ac.uk

ABSTRACT

Lundy is unique amongst British islands in having plants and insects that are known from nowhere else. How the ancestors of Lundy cabbage and its beetles may have come to be on Lundy is largely a mystery. They must have colonised Lundy sometime after the last Ice Age, at which time rising sea levels may not yet have turned it into an island. Lundy cabbage appears to have common ancestry with a closely related species including population(s) around the Bristol Channel, but the origins of the beetles are so far unclear and subject to current research. In recent times, numbers of Lundy cabbage have fluctuated greatly, probably in response to changes in rabbit abundance, but its range on Lundy is much less variable. Careful management, particularly of grazing animals and invasive rhododendron, is needed to ensure that this unique community continues to flourish.

Keywords: *Lundy cabbage, BAP, endemic, phylogeography, rabbit, rhododendron, sea-levels*

INTRODUCTION

Lundy is Britain's only offshore island that has its own endemic plant species with endemic insects feeding on it (Compton *et al.*, 2002). Reflecting this, the plant and its insects are listed on the United Kingdom Biodiversity Action Plan and have conservation action plans (UK BAP, 2001, Compton and Key, 1998), and have been the subject of conservation-related studies supported by Natural England (previously called English Nature) and others (Key *et al.*, 2000). We have been monitoring the plant and its insects since 1993, studying various aspects of their conservation ecology and evolutionary background, and developing plans to manage various aspects of their habitat to ensure its future on Lundy.

Here, after briefly describing the natural history of the species, we look at how such an interesting plant and insect community could have come to be present on Lundy, look at their distribution on Lundy and how numbers of plants have fluctuated in recent years, before speculating on the future for the plant and its insects.

THE PLANT AND ITS INSECTS

Lundy cabbage *Coincya wrightii* (Figure 1, bottom) is a short-lived perennial ruderal or pioneer member of the Brassicaceae, inhabiting sparsely vegetated rock on sea cliffs and is a quick coloniser of bare soil after disturbance, but is a poor competitor against regenerating grass swards (Compton and Key, 2000). It is restricted to the cliffs, some of the Sidelands (the steep, usually grassy slopes between the top of the sea cliffs and flatter area of the plateau) and rock outcrops (Buttresses) of the south-eastern coast of Lundy, from around Marisco Castle northwards to the Knights Templar Rock.

We have found that most of the insects that feed generally on other plants in the cabbage family also feed on Lundy cabbage, for example caterpillars of the large, small and green-veined white butterflies *Pieris brassicae, P. rapae* and *P. napi* occasionally defoliate individual plants, and a large number of insects of many orders have been found feeding on it (Compton & Key, 2000). Of particular interest are three beetles, two of which seem to be entirely restricted to Lundy. Best known is the 'bronze Lundy cabbage flea beetle' *Psylliodes luridipennis* (Chrysomelidae), (Figure 2, top left) whose larvae produce mines in the leaf petioles and stems. This has been shown to be a distinct species (Craven, 2002) that has so far never been found elsewhere. In the same genus, the 'blue Lundy cabbage flea beetle' has similar biology and appears to be a short-winged form of the widespread species *Psylliodes napi*, (Figure 2 top right) which on Lundy we have occasionally also found on bitter cress *Cardamine* spp. and Danish scurvy grass *Cochlearia danica*. Genetic studies are under way to ascertain the evolutionary relationship between populations from Lundy and other British and European populations of this beetle. It may be that this is just a widespread form of a beetle that happened to be found on Lundy first.

Loss of flight in insects is often said to be associated with life on islands (Roff, 1990), but we have recently found previously unrecorded populations of short-winged *P. napi* on Danish scurvy grass in North Devon and on other crucifers elsewhere in the U.K. and Europe, although always in mixed populations of long- and short-winged forms, unlike Lundy where all recorded individuals have had short wings. All individuals of *Psylliodes luridipennis* that we have investigated have been fully winged and we have often observed it in flight.

The third species, the 'Lundy cabbage weevil' (Figure 2, bottom), is also flightless and, from its pale yellowish/brownish legs in contrast to the black legs of the 'typical' form, is currently described as variety *pallipes* of the common cabbage leaf weevil *Ceutorhynchus contractus* (Curculionidae). (The precise nomenclature of *C. contractus* is currently under review and the name *C. minutus* has recently also been used (M.G. Morris, in prep.)). Larvae of this weevil mine the leaves of Lundy cabbage, and also Danish scurvy grass. Not all weevils of the species on Lundy have yellow legs, and the pale-legged form is far more common on Lundy cabbage than on Danish scurvy grass. We have found the distribution of leg colour morphs between the two host plants and between geographical locations on Lundy to be complex and baffling and are currently undertaking genetic studies to sort out the relationships between them and related beetles across Europe.

Figure 1: Isle of Man cabbage *Coincya monensis monensis* from Three Cliffs Bay, Gower (top left) and Wallasey dunes, Merseyside (top right) and Lundy cabbage *Coincya wrightii* from the cliffs above Landing Bay, Lundy (bottom)

Figure 2: Bronze Lundy cabbage flea beetle *Psylliodes luridipennis* (top left), blue Lundy cabbage flea beetle *Psylliodes napi* (Lundy form) (top right) and Lundy cabbage weevil *Ceutorhynchus contractus pallipes* (bottom). Scale bar = 1 mm

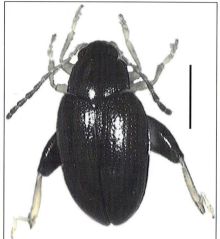

We have found all three species of beetle to occur throughout the range of their foodplant, occurring on plants growing in vertical sea cliffs, inland in Millcombe, and colonising plants seeded into experimental exclosures within the range of the foodplant in less than a year.

PAST

How did Lundy come to gain this unique community of plant and beetles? They or their ancestral forms could not have survived on Lundy through the last Ice Age, which at its maximum resulted in ice sheets that extended as far south as what is now the Bristol Channel. The area, including Lundy, will have experienced a tundra-like climate far too cold for these species. This means that Lundy cabbage and its beetles must either be relict species that were once more widespread (or that do occur elsewhere, but have yet to be discovered), or be 'new' taxa that have diverged *in situ* on or around Lundy, or, of course, the individual species of the assemblage may have different origins.

Together with Cinderella Grout, who was working on a Marie Curie fellowship at Leeds University, we have been using molecular genetic techniques to investigate the relationships between Lundy cabbage and other species of the genus *Coincya*, most of which are found in the Iberian Peninsula. Preliminary results suggest that Lundy cabbage is genetically very close to some populations of the Isle of Man cabbage *Coincya monensis monensis*, in particular to a single population of it growing on the Gower Peninsula in South Wales (Figure 1, top left), only 47km to the north-east. This population appears to be quite isolated genetically from other populations, currently referred as the same subspecies, in the rest of Britain and, together with *C. wrightii*, is actually closer to coastal populations considered to be *C. monensis cheiranthos* in Northern Spain (Cinderella Grout, unpublished data).

C. monensis is a very widespread, mainly annual species, with many named subspecies in Europe, especially in the Iberian Peninsula (Leadlay & Heywood; 1990). In the U.K. there are usually thought to be two subspecies. The Isle of Man cabbage *Coincya monensis monensis* (Figure 1 top) is endemic, associated with coastal dunes in the west, including the Isle of Man, and the Welsh, N.W. English and Scottish coasts. Subspecies *cheiranthos* (wallflower cabbage) is a ruderal plant of docklands and waste ground, fairly widely distributed but particularly so in South Wales, where it is spreading (Preston *et al.*, 2002). It is considered to be a recent introduction from mainland Europe where it is very widespread. It is unfortunate that the populations of *C. monensis monensis* recorded from the South Devon and Cornish coasts, always considered to be casuals (Stace, 1997) and last seen in the early twentieth century (National Biodiversity Network data), and those thought to be *C. monensis cheiranthos* from North Devon from the 1950s or '60s are now extinct and no genetic material remains to include in our study.

How might the Lundy cabbage - or its predecessor - have arrived on Lundy and maybe changed to become the plant we now call Lundy cabbage? To examine this we needed to be able to follow the history of Lundy from the end of the last Ice Age to when it eventually become habitable for the plants and animals present today. In particular it was important to establish at what point rising sea levels turned Lundy into an island and what were the likely climatic and ecological conditions at that time.

The reconstruction and dating of past environments is a complex problem, needing to take into account uplifting of the Earth's crust as the weight of ice reduced, the counteracting rise in sea level due to glacial meltwater and subsequent effects of erosion and deposition of sediments.

We examined how and when Lundy became an island by combining existing estimates of post-glacial sea level rise with the present day topography of Lundy and North Devon and the bathymetry of this area of the Bristol Channel, and then linked this to what is known of past climates to infer what conditions may have been like on Lundy at the time (Craven, 2002). To do this we obtained depth data for the area of the Bristol Channel around Lundy from Admiralty charts using *Leadline*, a

- 165 -

geographical information system tool designed by Tony Pilkington of the Geographic Information Unit at English Nature for the programme MapInfo. 'Artefacts' such as ship wrecks were digitally removed, as well as recently deposited sand banks including the two 'Banner Banks' to the N.E. and N.W. of Lundy which are the result of scour and deposition by currents of the modern Bristol Channel (Figure 3) (Stride, 1982). These banks were 'flattened' by taking the depths at their bases and replacing the values over the banks with a uniform value (-30 m). Finally we combined depth contours with terrestrial ones from the Ordnance Survey 1:50000 dataset and converted them all to a triangular irregular network (wire-frame model) using Intergraph Terrain Analysis, and used the MapInfo programme to generate a series of maps showing sea level rise at 2m intervals (Figure 4).

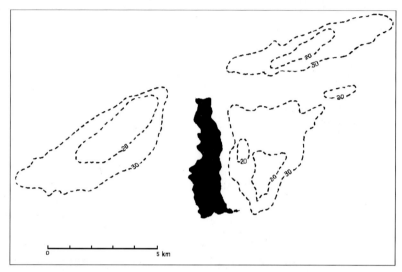

Figure 3: Map showing the sea depths surrounding Lundy in metres plotted from Admiralty Charts. The 'Banner Banks' to the north-east and north-west are recent accumulations of sediment and were excluded for modelling purposes

Using a sea-level curve developed by Lambeck for an area in the Celtic Sea located just to the west of Lundy, we added a time-scale for the sea level rise which allowed us to infer climatic conditions. The majority of climate data used in this work were amalgamated by the Quaternary Environments Network (Adams, 1997), based on pollen, fossil insect and past lake level data (Anderson, 1997; Atkinson et al., 1987; Harrison et al., 1996 respectively).

At the last glacial maximum, 25,500 years ago (Eyles & McCabe, 1989), a tongue of ice probably extended southwards into the Celtic Sea as far as the Scilly Isles (Scourse et al., 1990), although the main margin of the ice sheet did not reach as far south as the southwest of England (Jones & Keen, 1993; Doody, 1996). When the ice melted, the shoreline is believed to have remained stationary until about 14,000 years ago because crustal rebound matched sea level rise (Lambeck, 1995). At this time the coastline was still well to the west of Lundy (Figure 4: -58m).

Figure 4: Modelled changes in Lundy and North Devon shorelines, produced at 2m depth intervals with dates in 'real' years, changing in response to sea level rise since the last glaciation. Present day coastline dotted. Fine detail (lakes/small islands/ fine coastal detail etc.) are likely to be artefacts of the modelling process and are best ignored. Additional information: (a) middle of the Older Dryas cold period with subsequent warming of conditions; (b) the Younger Dryas ends marking the start of the Holocene; (c) the climate continues to warm but conditions are still cooler than present day; (d) climate begins to warm slightly. Sea = grey. Land = black

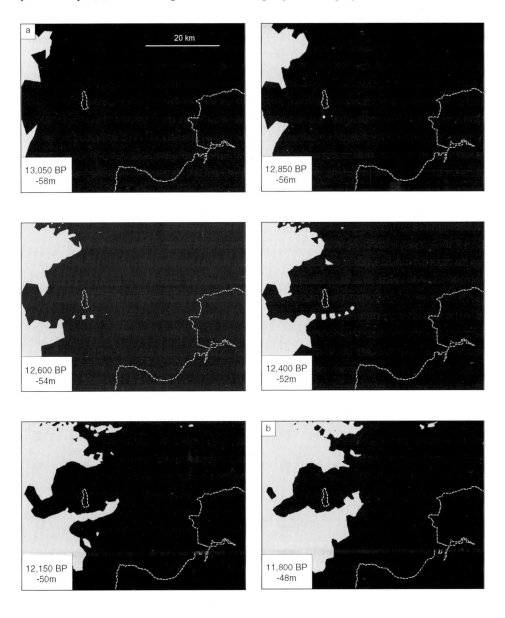

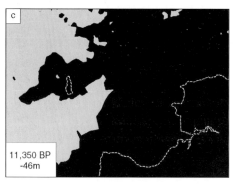

11,350 BP
-46m

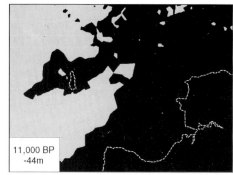

11,000 BP
-44m

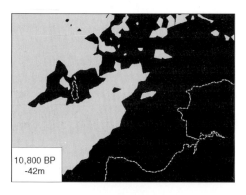

10,800 BP
-42m

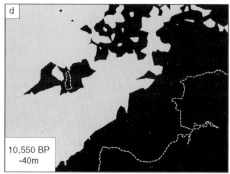

10,550 BP
-40m

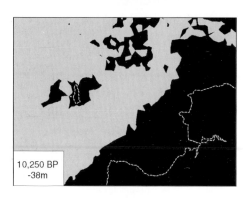

10,250 BP
-38m

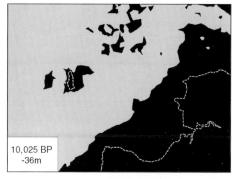

10,025 BP
-36m

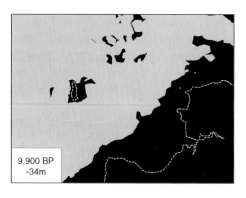

9,900 BP
-34m

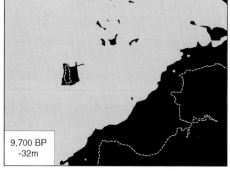

9,700 BP
-32m

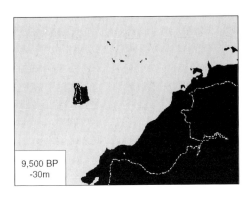

9,500 BP
-30m

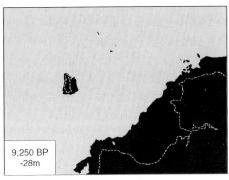

9,250 BP
-28m

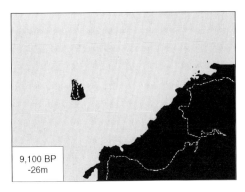

9,100 BP
-26m

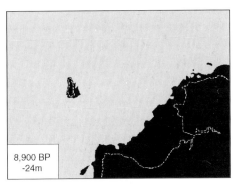

8,900 BP
-24m

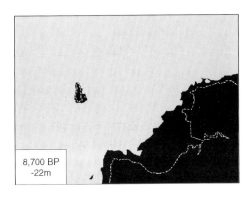

8,700 BP
-22m

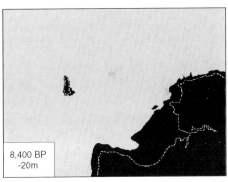

8,400 BP
-20m

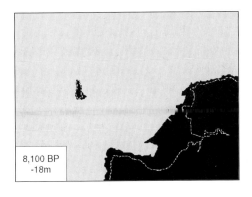

8,100 BP
-18m

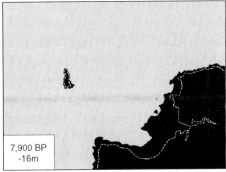

7,900 BP
-16m

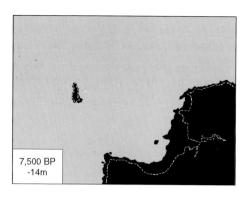

7,500 BP
-14m

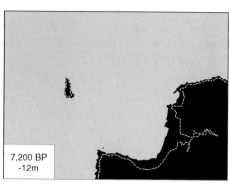

7,200 BP
-12m

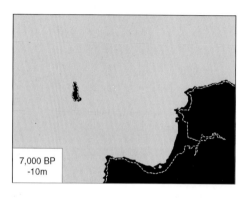

7,000 BP
-10m

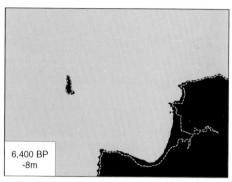

6,400 BP
-8m

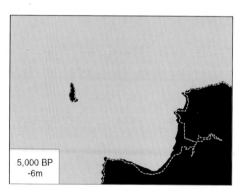

5,000 BP
-6m

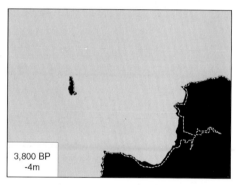

3,800 BP
-4m

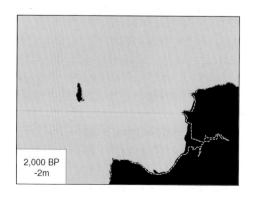

2,000 BP
-2m

Present
sea level

The climate remained very cold and dry after the glacial maximum, but shortly before 14,500 years ago it temporarily grew warmer and moister (Adams, 1997), but was followed by two periods of much colder conditions (the Older and Younger Dryas) which ended about 11,500 years ago when the current warm period began. From this time onwards climatic conditions started to become suitable for Lundy cabbage and its insects in the area.

Our model suggests that the widening estuary of the proto-Taw/Torridge (Figure 4: -52m to -54m) to the south of 'Lundy', then an isolated large granite hill or tor, eventually left 'Lundy' on a peninsula by the end of the Younger Dryas (Figure 4: -50m to -44m) extending southwest from what is now the middle of the Bristol Channel. This closely corresponds to the map produced by Gardner (1968), although our dating is somewhat different. As sea level rose, this spit reduced in area and its 'neck' narrowed until Lundy became an island (Figure 4: -42 to -38m) between 10,800 and 10,550 years ago, at a point when the climate was probably slightly cooler than now, but beginning to warm (Adams, 1997), and when conditions in southwest Britain were particularly mild.

Previous estimates of when Lundy became an island vary considerably (Table 1).

Table 1: Estimates of the date of isolation of Lundy as an island

Upper Palaeolithic 45,000-11,700 years ago	Schofield & Webster (1990)
10,800-10,550 years ago	This study
9,000 years ago	Gardner (1968)
>8900 years ago	Lambeck (1995)

Lambeck's (1995) calculations suggested that Lundy might have become isolated earlier than 8,900 years ago, but this model was acknowledged to have inadequate data for some areas, one of which was the Bristol Channel. As we used a sea level curve generated by a more recent version of Lambeck's model (personal communication from K. Lambeck to J. Scourse, 1999), coupled with detailed bathymetric data for the area around Lundy, our predictions of palaeoshorelines should be more accurate than has been possible before. Unfortunately, just as Gardner found in the 1960s, there remain insufficient data to estimate how much the present floor of the Bristol Channel has been altered by post-glacial sedimentation and erosion, and so the results of our model similarly must be treated with caution.

Over the next thousand years land to the northeast of Lundy slowly disappeared, but may have played an important role by providing 'stepping stones' for plants and animals colonising Lundy during this period (Figure 4: -40m to -32m). The distance between Lundy and North Devon increased as the coast receded towards its present day position (Figure 4: -30m to -2m).

From about 9,000 years ago, conditions were slightly warmer and moister than present, with another brief cool phase 8,100 years ago, though not as severe as the Younger Dryas (Adams, 1997). Temperatures rose to probably their highest since the Ice Age, between 7,900 and 4,500 years ago (Adams, *op. cit.*) whereupon the climate became largely similar to that of today.

Our results therefore suggest that the ancestors of Lundy cabbage and its beetles may have had the opportunity to colonise Lundy across land during a few hundred years around 10,800 years ago or may subsequently have been aided by 'stepping stone' land to the north east.

As Lundy became an island it would have been a high, flat-topped hill or tor standing out from flat plains and estuaries that are now beneath the sea. Widespread sand, including dunes, will have stretched along the proto-Taw/Torridge estuary, around Lundy northwards across a smaller Severn estuary to the Gower peninsula, probably providing wide expanses of habitat for dune plants such as the Isle of Man cabbage. One possible origin for the Lundy cabbage is that its progenitors, or a form ancestral to both it and the Gower (and Devon?) population(s) of *C. monensis monensis*, may have been isolated on sands around Lundy, possibly adapting to rockier conditions and persisting as the plant we know as Lundy cabbage today.

PRESENT

Our current studies of Lundy cabbage started in 1993 and since then we have been taking annual counts of the numbers of plants in flower across its whole range and, in a few accessible areas, also counting individuals that are not in flower (seedlings, non-flowering rosettes and plants prevented from flowering by herbivores). These data have provided insights into its distribution and fluctuations in abundance, and some of the factors that are influencing its population dynamics. Our views on the drivers influencing fluctuations in its numbers have changed considerably over this period, highlighting the value of such relatively long term data sets.

The overall range of Lundy cabbage has changed very little since 1993, in marked contrast to the numbers of plants that flower. It remains restricted to the coasts of the relatively sheltered south-eastern half of the island, extending inland only a couple of hundred metres in Millcombe. A combination of the effects of grazing and competition from other plants, notably grasses on deeper, moister soils on some of the Sidelands, determines its distribution. Few are found on the less steep areas within its range that are readily available to grazing animals, although in 'good' years, such as 2006, the plants colonise the grass and bracken-clad Sidelands just to the north of Millcombe and in the Marisco Castle fosse. The plants there rarely flower, however, and do not persist as they are grazed off. We have shown that a number of introduced grazing animals significantly impact on the Lundy cabbage; feral goats, Soay and domestic sheep and, in particular, rabbits.

The limiting role of grazing is evident on the cliffs and rock outcrops, where plants are restricted to the steepest sections inaccessible to goats and Soay sheep; on a small outcrop about 100m north of St Helena's Combe, which used to project above one of the main rhododendron patches, Lundy cabbage now only persists within a protective dense growth of bramble after clearance of the rhododendron.

Evidence for the role of competition from other vegetation comes from some of our experimental exclosures, where areas free of grazing animals temporarily supported Lundy cabbage. After a while, however, a dense grass sward developed and Lundy cabbage disappeared, especially on deeper, moister soils which support lush growth of Yorkshire fog *Holcus lanatus* and red fescue *Festuca rubra*. In such circumstances, disturbance from grazing animals may favour the cabbage. We failed in an experimental attempt to establish Lundy cabbage in an exclosure on one of the more grassy areas of the Sidelands just to the north of Hangman's Hill in 1997. Despite our initial clearance of grasses and other plants prior to seeding, the resultant Lundy cabbage seedlings were rapidly out-competed by grass regeneration.

We have shown that Lundy cabbage is also subject to competition (and indeed elimination in places) by the alien shrub *Rhododendron ponticum* (Compton *et al*, 1997, 1999) and the area of rock outcrop and cliff face available to the cabbage has been reduced since its rapid population expansion really started after a fire in 1926. We have shown that all populations of Lundy cabbage are vulnerable to supplanting by rhododendron and so it poses a continual threat as long as it remains on the island. Colonisation of cliffs by Lundy cabbage after clearance of rhododendron is, however, rapid and spectacular and the immediate threat that rhododendron poses has been reduced somewhat in recent years, thanks to the past and current efforts of the island's wardens, rangers, numerous volunteers, and the cliff-climbers of *Ropeworks*, led by Angus Tillotson.

Numbers of Lundy cabbage in flower have fluctuated considerably, but not erratically. A very distinct pattern has emerged of widely separated peaks and troughs in plant numbers varying up to ten-fold (Figures 5 and 6). We now consider that these changes in abundance are driven largely by the huge variations in the number of rabbits, driven by outbreaks of myxomatosis that have taken place over the same period (Compton *et al*., 2004). With almost no predators, rabbits thrive on Lundy, leading to very closely-cropped swards, bare ground and erosion. Myxomatosis first arrived on the island in 1983 and the numbers of rabbits crashed, as they have done three times subsequently. In response, the populations of many plant species, including Lundy cabbage, rapidly recover and flower profusely. Pioneer species such as the cabbage, which are favoured by bare ground and disturbance, respond particularly dramatically. This effect is relatively short lived, however, as rabbit numbers inevitably quickly recover, and regeneration of the cabbage is subsequently suppressed by a combination of competition with other plants followed by intense rabbit grazing.

We have found that fluctuations in numbers of Lundy's special beetles are intimately linked with the abundance of their host plant. In the early 2000s, when the Lundy cabbage was scarce, it was very hard to find its dependent insects and few plants appeared to have any of the unique species of beetle on them. The beetles' distribution became very patchy, with perhaps the highest numbers remaining on a small population of the cabbage near the beach at Quarry Bay. Cabbage numbers recovered rapidly in 2005 and 2006, but the recovery of its beetles seemed to be lagging behind. While numbers of all the beetles increased, they had very large numbers of plants to recolonise, making them more difficult to find. *Ceutorhynchus contractus*, including var. *pallipes*, seemed to recover much more rapidly than either *Psylliodes* species, perhaps because it has more generations in a year and has had its alternative food plant (scurvy grass) to fall back on. The winged *Psylliodes luridipennis*

Figure 5: Variation in Lundy cabbage numbers between years with low numbers of rabbits (2006 top) and high numbers (2005 bottom). Sideland population along beach road, just south of the turn into Millcombe

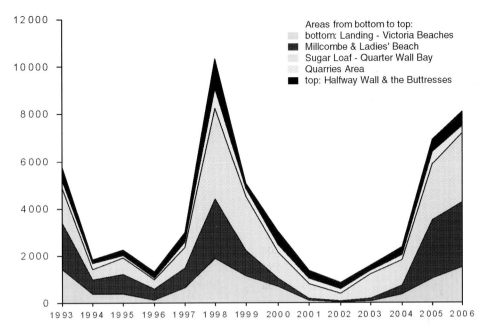

Figure 6: Number of plants of Lundy cabbage in flower in the main areas of its distribution between 1993-2006. Data from 1993 approximated from incomplete data by interpolation of numbers based on average proportion of total count contributed by each counting unit for all units in subsequent years

has seemed to be able to recolonise more quickly than the flightless *P. napi*. What effect such repeated population bottlenecks may be having or have had on the genetics of the beetles, we do not know.

FUTURE

The continuing survival of Lundy cabbage is more or less assured for the time being. Its numbers show no sign of long-term decline, while the Millennium Seedbank at the Royal Botanic Gardens at Kew has thousands of Lundy cabbage seeds which could be used in an emergency, although hopefully this is unlikely ever to be necessary. Awareness of its ecological needs has improved and has informed the management of the domestic sheep and larger feral mammals to accommodate the plants' needs and these seem now to pose less of a threat to the plant within in its current range than in the past.

Work on the rhododendron is by no means complete and more clearance is still needed, urgently in some areas, such as the cliffs above and to the south of Quarry Bay where it is still actively invading areas of Lundy cabbage. Continual clearance without the objective of its eradication is clearly unsustainable and, until this highly invasive species is entirely eliminated from Lundy it will remain a threat and could very quickly re-colonise all the areas so far cleared, undoing all the hard work and negating the financial resources that have gone into its control.

Whether Lundy cabbage numbers continue to fluctuate in the way we have seen in recent years is likely to be dependent mostly on future management of rabbit numbers. 'Boom and bust' cycles in rabbit numbers driven by myxomatosis will probably mean continuation of the cycles in the numbers of the plant and insects, with unpredictable implications for their genetic diversity. Rabbit numbers stabilised at a low level would probably result in fewer Lundy cabbage destroyed by grazing, with less disturbance and erosion and fewer areas of bare ground into which the cabbage may regenerate. Vegetation on the Sidelands could become even more dense and grassy, maintaining the absence of Lundy cabbage there. In such a scenario overall numbers of Lundy cabbage would be likely to be more stable, perhaps at a figure somewhere between the highs and lows of the recent past, and the dramatic 'shows' of the plant that were witnessed in the post myxomatosis years of 1993, 1998 and 2005-6 might not be seen again, at least on the Sidelands. For the insects, the implications of recent observations are that wildly fluctuating populations of Lundy cabbage may be more of a problem for them than for the plant itself.

Longer-term, it is by no means clear what climate change might have in store for the vegetation and wildlife of Lundy. Various conflicting models predict either considerable warming or cooling of the British climate, together with either reduced or enhanced precipitation in the west of Britain. However, the predicted climate scenarios for the next century or two are generally within the variation already seen during the post-glacial period and have probably already been experienced by Lundy cabbage and its fauna. Different climatic conditions will also influence the interplay between Lundy cabbage, competing vegetation, grazing animals and whatever additional species may colonise Lundy within the remainder of the current interglacial.

After that ... ?

ACKNOWLEDGEMENTS

Funding for our research on Lundy has been provided by English Nature, Leeds University, by the Natural Environment Research Council/English Nature for M.Res. and Ph.D. studentships for Jenny Craven and a European Union Marie Curie fellowship for Cinderella Grout, to whom thanks are due for the use of her unpublished data. The work was made possible (and pleasurable) by the help and kindness over the years from all of the Lundy wardens and other staff in their blue tee-shirts. Tony Pilkington in English Nature's Geographical Information System Unit helped considerably in detailing the post-glacial sea level changes around Lundy.

REFERENCES

Adams, J. 1997. Global land environments since the last interglacial: Europe during the last 150,000 years. Oak Ridge National Laboratory, TN, USA. *http://www.esd.ornl.gov/projects/qen/nercEUROPE.html* (Accessed 18.1.02).

Anderson, D.E. 1997. Younger Dryas research and its implications for understanding abrupt climatic change. *Progress in Physical Geography* 21, 230-249.

Atkinson, T.C., Briffa, K.R. & Coope, G.R. 1987. Seasonal temperatures in Britain during the past 22,000 years, reconstructed using beetle remains. *Nature*, 325, 587-592.

Compton, S.G. & Key, R.S. 1998. *Species Action Plan: Lundy Cabbage (Coincya wrightii) and its associated insects.* Unpublished report to Species Recovery Program, English Nature, Peterborough.

Compton, S.G. & Key, R.S. 2000. *Coincya wrightii* (O.E. Schulz) Stace. *Journal of Ecology*, 88, 535-547.

Compton, S.G., Key, R.S. & Key, R.J.D. 1999. *Rhododendron ponticum* on Lundy - beautiful but dangerous. *Annual Report of the Lundy Field Society 1998*, 49: 74-81.

Compton, S.G., Key, R.S. & Key, R.J.D. 2002. Conserving our little Galapagos - Lundy, Lundy cabbage and its beetles. *British Wildlife*, 13, 184-190.

Compton, S.G., Key, R.S. & Key, R.J.D. 2004. Lundy cabbage population peaks - are they driven by rabbits and myxomatosis? *Annual Report of the Lundy Field Society 2003*, 53, 50-56.

Compton, S.G. Key, R.S., Key, R.J.D. & Parkes E. 1997. Control of *Rhododendron ponticum* on Lundy in relation to the conservation of the endemic plant Lundy cabbage, *Coincya wrightii*. *English Nature Research Reports*, 263, 1-67.

Craven, J.C. 2002. *The Ecology and Evolution of the Bronze Lundy Cabbage Flea Beetle, Psylliodes luridipennis.* Master of Research Thesis. University of Leeds. 113pp.

Doody, J.P. 1996. Chapter 1.2 Introduction to the region. In *Coasts and Seas of the United Kingdom. Region 11 The Western Approaches: Falmouth Bay to Kenfig* (Eds. J.H. Barne, C.F. Robson, S.S. Kaznowska, J.P. Doody, N.C. Davidson, & A.L. Buck), Joint Nature Conservation Committee, Peterborough, (Coastal Directories Series) pp. 13-18.

Eyles, N. & McCabe, A.M. 1989. The Late Devensian (<22,000 BP) Irish Sea basin: the sedimentary record of a collapsed ice sheet margin. *Quaternary Science Reviews*, 8, 307-351.

Gardner, K. 1968. Lundy - a Mesolithic peninsula? *Annual Report of the Lundy Field Society 1967*, 18, 24-28.

Harrison, S.P., Yu, G. & Tarasov, P.E. 1996. Late Quaternary lake-level record from northern Eurasia. *Quaternary Research*, 45, 138-159.

Jones R.L. & Keen D.H. 1993. *Pleistocene Environments in the British Isles.* London: Chapman and Hall.

Key, R.S., Compton, S.G. & Key, R.J.D. 2000. Conservation studies of the Lundy cabbage between 1994 and 2000. *Annual Report of the Lundy Field Society 1999*, 49, 74-81.

Lambeck, K. 1995. Late Devensian and Holocene shorelines of the British Isles and North Sea from models of glacio-hydro-isostatic rebound. *Journal of the Geological Society of London*, 152, 437-448.

Leadlay, E.A. & Heywood, V.H. 1990. The biology and systematics of the genus *Coincya* Porta & Rigo ex Rouy (Cruciferae). *Botanical Journal of the Linnaean Society*, 102, 313-398.

Morris, M.G. (in prep.) *Handbooks for the Identification of British Insects. True Weevils* 5(17c). Coleoptera : Curculionidae (Subfamily Ceutorhynchinae). Royal Entomological Society/Field Studies Council.

Preston, C.D., Pearman, D.A. & Dines, T.D. 2002. *New Atlas of the British and Irish Flora*. Oxford: Oxford University Press.

Richardson, S.J., Compton, S.G. & Whitely, G.M. 1998. Run-off of fertiliser nitrate on Lundy and its potential ecological consequences. *Annual Report of the Lundy Field Society 1997*, 48, 94-102.

Roff, D.A. 1990. The evolution of flightlessness in insects. *Ecological Monographs*, 60, 389-421.

Schofield, A.J. & Webster, C.J. 1990. Archaeological Fieldwork 1989. Further Test Pit Excavations South of Quarter Wall. *Annual Report of the Lundy Field Society 1989*, 40, 34-47.

Scourse, J.D., Bateman, R.M., Catt, J.A., Evans C.D.R., Robinson, J.E. & Young, J.R. 1990. Sedimentology and micropalaeontology of glacimarine sediments from the central and south-western Celtic Sea. In *Glacimarine Sediments: Processes and Sediments* (Eds. J.A. Dowdeswell & J.D. Scourse), Geological Society, London. Special Publication 53, 329-347.

Stace, C. 1997. *New flora of the British Isles. Second Edition*. Cambridge: Cambridge University Press.

UK BAP. 2001 Species Action Plan Lundy Cabbage (*Coincya wrightii*) *http://www.ukbap.org.uk/UKPlans.aspx?ID=232*

TERRESTRIAL ECOLOGY: DISCUSSION

(Initials: SC=Stephen Compton, RK=Roger Key, JH=John Hedger, TP=Tony Parsons,
JM=John Morgan, MT=Myrtle Ternstrom, Q=Unknown participant)

Q: *You mentioned, Dr Compton, that Lundy has several things in common with St Kilda. There are Soay sheep on St Kilda also. When Soay sheep were introduced to Lundy, could they have carried seeds of the cabbage plant with them?*

SC: The cabbage plant is not on St Kilda, only the one beetle. The beetle could have moved, but it is now regarded as a separate species.

Q: *When was the cabbage first referred to on Lundy? Did it pre-date the introduction of the Soay sheep?*

SC: The Soay sheep were introduced in the 1940s. The cabbage or its ancestor has probably been on Lundy for thousands of years.

Q: *Does the cabbage occur elsewhere on the Gower peninsula other than the place you mention?*

SC: We did our sampling at Three Cliffs Bay on the Gower and here the cabbage is *Coincya monensis monensis*, which also occurs on the Isle of Man. The population of this subspecies on the Gower is genetically close to the Lundy cabbage, *Coincya wrightii*.

RK: If you look in the British Atlas there are a number of places where *Coincya monensis* occurs along the South Wales coast. All, except the ones in Three Cliffs Bay are the wallflower cabbage, a different subspecies, *Coincya monensis cheiranthos*, a Pan-European subspecies which has spread to North America where it has become a pest.

Q: *Commercial sheep numbers on Lundy have decreased since 2003 and the feral life e.g. deer, has become carefully managed. You said rabbits influence cabbage numbers. Reduction in sheep will mean fewer rabbits as the sheep make the habitat suitable for rabbits which do not like long sward heights. If you take out the years when myxomatosis occurred, there is an inverse correlation between the number of sheep and the number of rabbits. At present with the removal of sheep, rabbits are increasing, but this will not last and it will go the other way.*

SC: We have done counts on the numbers of plants in various areas and this has given insight into fluctuations in abundance. These changes appear to be driven largely by the large variation in rabbit numbers that occurs.

Q: *If you remove sheep and there is not so much grazing, will this affect the growth of the fungi as well as the cabbage?*

SC: As far as the cabbage is concerned there is spatial separation as the sheep are kept away from the cabbage areas.

JH: This is an example of a conservation dilemma - the assistance of the survival of one species at the expense of another! The best areas for the fungi on Lundy are on the centre of the plateau of the island, especially the Airfield and Middle Park. The Sidelands with their good plant diversity, where the cabbage lives, are good for pathogenic fungi. There is really no conflict as good management will reduce the excessive grazing that causes problems such as erosion. Plant biodiversity increases with grazing. If you stop grazing as Stephen has done in his exclosure experiments, then the Yorkshire fog, *Holcus lanatus*, and other grasses will take over. It is the heavy grazing and lack of fertilizer input that makes Lundy so exciting botanically. The most boring area for fungi on Lundy is the improved grassland where you only find one species, *Hygrocybe virginea*, which is tolerant of high nitrogen levels.

TP: It is important to remember that the species we are protecting now have been on the island for thousands of years. It is only in the last 40-50 years that the human species has been causing problems. If we return to subsistence farming on Lundy, e.g. farmed rabbits that are taken over to the mainland for sale, limited sheep, no fertilizers, then the original ecological balance may return, but I am not saying that we should; however this could be the reason why we have got the existing species on Lundy today.

JM: *My question concerns the behaviour of the bluebell, which I thought was a woodland plant. In recent years in May it has moved against the prevailing wind over the top of the island, where its stalks have become much shorter.*

TP: In Normandy and parts of Dorset you find stands of bluebells all down the cliffs. It is not just a woodland plant.

JH: Lundy can be compared with the similar islands of Skomer and Skokholm where you find bluebells, especially on Skomer, covering the island to the edges of the cliffs. It is also interesting to note that there is a fungus associated with the bluebell, *Uromyces muscari.*

Q: *A comment on the last question firstly. You find sheets of bluebells in open country from High Tor on the top of Dartmoor. A more maritime climate brings the species out from cover. My question is: was Lundy heavily wooded in the past?*

JH: Hubbard in her paper on trees in the LFS 21st Annual Report says that Lundy probably became deforested by the thirteenth century. This was due to usage of wood by the inhabitants and also to the sailing ships that came to Lundy for repairs. Obviously some fungi became extinct with the removal of the trees but the wood-decomposer fungi will come back in. Lundy must have been forested in the Boreal period around 5000 B.C.

MT: *Surely the destruction of the rhododendron will affect the birds and the deer will lose their habitat?*

TP: There has to be a balance. If there is a problem with the breeding land birds it can be solved by the planting of native trees and this will help the deer also.

POSTER ABSTRACTS

SIXTY YEARS OF THE LUNDY FIELD SOCIETY
by
KATE COLE

The Flat, Woldringfold, Burnt House Lane, Lower Beeding, West Sussex, RH13 6NL
e-mail: kate.cole@btopenworld.com

The Lundy Field Society (LFS) was founded in 1946, following discussions between Leslie Harvey, a lecturer in Zoology at the University College of the South West, and Martin Coles Harman, the owner of the Island and a keen naturalist. The original aim of the Society was to establish Lundy as a bird observatory with proposals for work including the erection of a Heligoland trap, the presence of biologists to operate the trap, ringing of nesting cliff-breeding birds, and periodic publication of progress reports and results. Harman was keen to support the project, as long as Lundy's independence from the mainland was maintained. Harman became the first President of the LFS and offered accommodation in the hotel and subsequently the Old Light which remained the headquarters of the Society until 1968. The Society's first year was spent organising the construction of the trap and maintaining the Old Light, with the first warden being appointed in 1947. Harman suggested a long-term project to repoint Marisco Castle, and this broadening of the LFS's interests is reflected in the first Annual Report which, as well as birds, contained preliminary reports on terrestrial and freshwater habitats and marine ecology. Sixty years on, the LFS still works to further the study of Lundy, and in particular its history, natural history and archaeology, and to conserve its wildlife and antiquities. Thanks in part to the work of the Society, much of the Island is now legally protected and a full-time warden is employed by the Landmark Trust and Natural England. Consequently, the LFS now plays a less direct role in conservation, but continues to supply volunteers to assist the warden in conservation tasks. The LFS also offers grants to encourage scientific research, with the results being published in an Annual Report.

THE BIRDS OF LUNDY
by
RICHARD CASTLE

The Pipits, 91 Maney Hill Road, Sutton Coldfield, B72 1JT
e-mail: richfran.castle@blueyonder.co.uk

Our knowledge of the birds that both breed on and visit the Island has been increasing steadily ever since the earliest written records were created in the early 1870s. It particularly improved after the foundation of a Bird Observatory in 1947 and the employment of full-time Resident Wardens via funding from the Lundy Field Society.

This display shows a selection of both the breeding birds found on Lundy and the migrants that visit the Island. Peter Davis (Resident Warden), in his 1954 book put the number of species seen on Lundy at 218. This rose to 274 when J.N. Dymond compiled the second 'The Birds of Lundy' in 1980. The latest definitive update is currently being compiled with the latest total being 318 (332 if you include unsuccessful introductions, etc).

For the breeding birds we are at an exciting time, as we have already seen positive effects from the recent eradication of rats from the island (both puffins and Manx shearwaters have now successfully bred for the first time in many years). Other nesting birds seem to have also increased in numbers, e.g. wheatears and stonechats, although they would still be under predation pressure from the gulls and crows.

For many visiting bird watchers there is always the excitement of seeing migration, especially during the main migration periods of April/May and August through to November. There is also the possibly of seeing rare species that do not normally visit our shores. The island's location results in birds from both America and the Far East landing to rest before moving on to the mainland.

Please could all visitors to the island enter any bird watching records into the LFS log, which is in the Tavern. The more we know about the birds that breed on and visit the Island the more we can help them.

THE LUNDY SEABIRD RECOVERY PROJECT: A BRIGHTER FUTURE FOR LUNDY'S BURROW-NESTING SEABIRDS

by

DAVID APPLETON[1], HELEN BOOKER[2], DAVID J. BULLOCK[3], LUCY CORDREY[3]
and BEN SAMPSON[4]

[1] Natural England, Level 2 Renslade House, Bonhay Road, Exeter, Devon, EX4 3AW
[2] RSPB, Keble House, Southernhay Gardens, Exeter, Devon, EX1 1NT
[3] The National Trust, Heelis, Kemble Drive, Swindon, SN2 2NA
[4] The Landmark Trust, Lundy Island, Bristol Channel, Devon, EX39 2LY

The U.K. holds over 90% of the global breeding population of Manx shearwater *Puffinus puffinus*. Lundy Island's population of Manx shearwater (and also puffin *Fratercula arctica*) is much lower than those reported in the mid-twentieth century. The impact of rats on seabird populations has been globally well documented. A major factor affecting the burrow-nesting species on Lundy was believed to be predation by black (*Rattus rattus*) and brown (*R. norvegicus*) rats. Both species are globally widespread and abundant and neither is native in the U.K. A partnership was formed to implement the Lundy Seabird Recovery Project, the primary objective of which was to eradicate rats to increase seabird breeding success. The project was controversial because, in the U.K., the black rat is rare. Between November 2002 and March 2004, the eradication programme was implemented. Following a further two years of checks, Lundy was declared rat-free in March 2006. Monitoring now focuses on the productivity and population trends of the target seabirds. Post-eradication estimates of Manx shearwater productivity are encouraging. However, it will be at least five years before these juveniles, the first recorded for c.50 years, return to Lundy and boost the breeding population.

TERRESTRIAL MAMMALS ON LUNDY:
AN ODD, RICH AND DYNAMIC ASSEMBLAGE
by
DAVID J. BULLOCK and LUCY CORDREY

The National Trust, Heelis, Kemble Drive, Swindon, SN2 2NA

Twenty-two terrestrial mammal species have been recorded on Lundy, many more than most other islands of similar size and isolation. However, perhaps only three (all bats) are truly native. Of these only one (common pipistrelle) appears to be resident. The remainder are: (1) early accidental introductions (mice, rats, a shrew); (2) livestock (cattle, sheep, goat, pig), horse/pony, dog and cat, some of which were established as feral populations (Soay sheep) or have become so (goat); (3) an eclectic mix introduced mainly in the early part of the last century (three deer, a marsupial, brown hare and red squirrel) of which only sika deer persists today. Rats (both species) were eradicated early this century to benefit burrow-nesting seabirds. Today the high combined biomass of sika, commercial sheep, feral sheep and goats, and rabbits has a major impact on Lundy's vegetation. In order to reduce the grazing pressure on features of nature conservation interest in the SSSI (lowland heath and coastal grasslands) the number of commercial sheep has recently been lowered as part of an agri-environment scheme. Annual counts by the LFS and our teams reveal that the reduction in commercial sheep has not been accompanied by increases in the other large herbivores, the numbers of which are stabilised by annual culls or live sales. However, as commercial sheep are reduced the number of rabbits has tended to increase suggesting competition for forage. The rabbit population, usually high, damages historic sites, buildings, farming interests and nature conservation and has proved very difficult to reduce. The current (2006) rabbit population is very low due to a myxomatosis outbreak. In the absence of culling now and into the winter it is likely to increase to pre-myxomatosis levels. We recommend continued monitoring of the large herbivores and rabbits on Lundy.

LICHENS OF LUNDY

by

PETER JAMES[1], ANN ALLEN[2] and BARBARA HILTON[2]

[1]19 Edith Road, London, W14 0SU
[2]Beauregard, 5 Alscott Gardens, Alverdiscott, Barnstaple, Devon, EX31 3QJ

Lichens are very special on Lundy; over 350 different species have been recorded, one-fifth of the total lichens of Great Britain - an amazing diversity for such a small island.

Lundy owes its lichen diversity to its south-western position and variety of relatively undisturbed and unpolluted habitats.

Lundy is an important reservoir of oceanic lichens, including a number of rare lichens (e.g. *Teloschistes flavicans* and *Anaptychia ciliaris* subsp. *mamillata*) and granite domes at the north which show a unique lichen succession sequence.

The largest number of lichen species is associated with siliceous granite (and to a lesser extent slate) of natural cliffs, maritime outcrops and boulders.

Other distinctive saxicolous communities are found on:

1. dry stone walls (e.g. Halfway Wall running east-west) showing effects of aspect on lichen distribution;
2. mineral-deficient rock and damp walls of the quarries;
3. old gravestones in Beacon Hill Cemetery, which includes marble, granite and slate memorials;
4. nutrient-enriched granite domes (e.g. in heathland along the west and at the north);
5. standing stones (acting as bird perches) along the track to the north;
6. mortar and cement in man-made buildings, ruins (e.g. the Battery) and walls which introduce a calcareous substrate.

Lundy boasts 120 different corticolous lichen species with 56 species recorded on *Acer pseudoplatanus*. With but few trees (concentrated in the Millcombe valley and around the quarries on the east) this is remarkable. Many trees are old, suffer from wind damage and require continuing sympathetic replacement.

Heathland is a great strength of Lundy; the wet area around Pondsbury is rich in *Cladonia* species on peat and heather. This, and the developing lichen community to the north, are fragile, sensitive to trampling, over-grazing, drought conditions and invasion by bracken. This outstanding community requires careful monitoring.

MONITORING THE LUNDY NO-TAKE ZONE - THE FIRST THREE YEARS

by

MILES HOSKIN[1], CHRIS DAVIS[2], ROSS COLEMAN[3] and KEITH HISCOCK[4]

[1] Coastal & Marine Environmental Research, 2 Raleigh Place, Falmouth, TR11 3QJ
[2] Natural England, Renslade House, Bonhay Road, Exeter, EX4 3AW
[3] Marine Ecology Laboratories, University of Sydney, NSW 2006, Australia
[4] Marine Biological Association, Citadel Hill, Plymouth, PL1 2PB
Corresponding author, e-mail: miles.hoskin@cmer.co.uk

The Lundy No-Take Zone (NTZ) is a 3.3 km^2 area off the east coast of the island, bounded to the east by the 4° 39'E line of latitude, but excluding the area of the Landing Bay. It is the U.K.'s first statutory NTZ for nature conservation.

Three aspects of Lundy's marine biodiversity have been selected for monitoring potential effects of the Lundy NTZ:

1. lobsters and crabs;
2. scallops, and
3. long-lived epifauna on subtidal rocks.

Lobsters and crabs are assessed for abundance and the sizes of individuals by experimental potting at locations within the NTZ, at nearby control sites outside the NTZ and at more-distant reference sites in North Devon and South-West Wales. Scallop density and size are assessed by diver measurement at sample locations within and outside of the NTZ on the east coast of Lundy. Epifauna on subtidal rocks are monitored by quadrat sampling (counts by divers) at two locations within the NTZ and two outside. Twenty-one species are included in the monitoring.

Monitoring has been undertaken in 2004, 2005 and 2006. Some of the results already show significant differences within and outside of the NTZ. The abundance and size of lobsters has been much greater within the NTZ than outside and the size of scallops larger inside than outside. Epifauna communities, because of their slow growth and often great longevity, will be much longer to show any effect. The present programme of monitoring will finish in 2007, but it is expected to be repeated again in the future to assess longer-term changes.

CONTRIBUTOR PROFILES

SHIRLEY BLAYLOCK

Shirley Blaylock is the National Trust Archaeologist for Devon and Cornwall. She first visited Lundy in 1993 and has worked for the National Trust in Devon since 1994, joining the archaeological survey team for seasons on Lundy from 1996. She is a joint author, with Caroline Thackray, of the current National Trust archaeological field guide to Lundy.

ROGER CHAPPLE

Roger Chapple first visited Lundy by Campbell steamer in the 1950s and later stayed at Signal Cottages in the mid 1970s. After much encouragement he joined the Lundy Field Society in 1997 and considers it a very great privilege to have been elected Chairman of the Society in 2002. He enjoyed the good fortune of attending the local Grammar School in Barnstaple at the same time as Denver Scoins who, after a career deep sea, was appointed first as Master of the *Polar Bear* and then became the Master of M.S. *Oldenburg*. Roger is a member of several local and national organisations and enjoys sport, gardening and walking. He is a member of local theatrical groups and regularly attends the activities of the Morgan Sports Car Club. Roger runs his own construction company in North Devon. He is married to Paula and they have four children.

DR STEPHEN COMPTON

After lecturing in South Africa for a few years Stephen returned to Yorkshire, and at present he is Reader in Entomology at Leeds University. His BSc degree was in Zoology at Hull University, where he stayed to do a PhD on cyanide polymorphism in birdsfoot trefoil. His interest in wildlife has been life-long. He was one of those kids who was rearing butterflies from caterpillars before they went to school and spilling frog spawn all over the back steps when not much older. Working as a professional biologist has allowed him to travel widely, from glaciers in Norway to the volcano of Krakatoa and the rainforests of Borneo. This has mainly been in connection with his main research interest - the ecology of fig trees and their pollinators. Closer to home, he has been involved with the conservation of several rare plant-feeding beetles.

JENNY CRAVEN

Jenny has always been a keen naturalist and she graduated from the University of Leeds in 2001 with a degree in Biology, which had included a year studying agronomy in a French university, and a dissertation on spider community-ecology in farmland. She spent a summer working as a research assistant at the Natural History Museum on the taxonomy of mites living on British bats, and, pursuing the invertebrate ecology theme, she subsequently went on to do a Masters by Research (MRes) in Biodiversity and Conservation at the University of Leeds, studying the ecology and evolution of the Lundy cabbage beetles. She wanted to continue the research she had started during her MRes, so she embarked on a PhD funded by NERC with Natural England as CASE partners. Jenny is supervised by Stephen Compton and Roger Key, and also Roger Butlin at the University of Sheffield, and she is now approaching the end of her PhD research.

DR LEWIS DEACON

Dr Lewis Deacon graduated in Biology at Portsmouth University and subsequently studied for a PhD entitled 'Functional biodiversity of grassland saprotrophic fungi' at King's College London as part of the NERC Soil Biodiversity thematic programme. He joined NSRI in September 2004, and is currently working on a BBSRC funded post-doctoral position entitled 'Self-organisation in the soil: microbe complex' collaborating with the University of Abertay, Dundee. He is a member of the British Mycological Society, the Institute of Biology and The British Society for Soil Science. Areas of expertise are in soil microbial ecology, analysis and characterisation of microbial communities for structure and function, Basidiomycete identification and surveying.

DR DAVID GEORGE

David George first visited Lundy in 1971 when he dived along with his wife Jenny George on a marine expedition led by Keith Hiscock to document its rich underwater life. Since that time he has dived around Lundy on many occasions on behalf of the Natural History Museum, London, and published on aspects of its marine invertebrate fauna in the LFS Annual Report. In recent years his diving has been largely confined to the warmer waters of the tropics and his Lundy activities have centred on the island's freshwater and terrestrial ecology, helping his wife with her pond life investigations and John Hedger with his detailed surveys of the island's fungi.

PROFESSOR JENNIFER GEORGE

Jennifer George is currently a Vice-President of the Lundy Field Society and was Chairman from 1988 to 2002. She has carried out research on the Lundy freshwater ecosystems since the late 1970s, and has published her results in the LFS Annual Reports. With over 30 peer-reviewed research papers and joint authorship (with David George) of an encyclopaedia of marine invertebrates she gained her Professoriate from the University of Westminster in 1990. Upon her retirement from that University in August 2003 where she was Provost of the Science and Technology campus, she was conferred Professor Emeritus by the University and is now involved with research, consultancy and committee work.

DR GARETH WYN GRIFFITH

Gareth Griffith is a lecturer in Mycology at the University of Wales Aberystwyth (UWA), specialising in fungal ecology. As an undergraduate at UWA he studied Microbiology, followed by a PhD in tropical fungal ecology also at UWA. Following periods of postdoctoral research at Glasgow and Bangor, he returned to UWA in

1996, where he has since focused on the ecology and conservation of grassland basidiomycetes, notably waxcaps, and several publications have ensued. With funding from NERC and statutory conservation bodies his research group has used GIS mapping of long-term field sites, stable isotope markers and genetic analyses to elucidate the biology of these fungi.

PROFESSOR JOHN HEDGER

John was born in Horsham, W. Sussex in 1945 where his early interest in Natural History, most especially fungi, was encouraged by joining the Horsham Natural History Society at the age of 8. He attended Collyer's School, Horsham, prior to reading Botany at Pembroke College, Cambridge from 1964-1967, followed by a PhD in Mycology in the Botany School, Cambridge, from 1967 to 1970, when he became a lecturer in Mycology at the University of Wales, Aberystwyth. He moved to the School of Biosciences, University of Westminster, as a Quintin Hogg Research Fellow, in 1996 and became Professor of Tropical Mycology. His mycological research interests include the ecology of fungi, most especially in Tropical Rainforest, and he has worked extensively in Indonesia, Papua New Guinea and Ecuador. He visited Lundy for the first time in 2003 to assist with a field study by Professor Jenny George, and this led to the publication, with collaborator David George (Natural History Museum), of a preliminary survey of the fungi on Lundy, in the 2003 Annual Report of the LFS. He is a member of the LFS.

DR KEITH HISCOCK

Keith Hiscock was born and brought-up in Ilfracombe so that Lundy was more accessible to him than for most. After becoming fascinated by seashore life, he took a zoology degree and during that time dived on Lundy for the first time in 1969. That trip provided a glimpse of the outstanding quality and variety of marine wildlife around Lundy and, as a highlight, he discovered colonies of the sunset cup coral there: for the first time in Britain. Over the next twenty years, he returned to the island with a wide range of colleagues to document and better understand the marine ecology of Lundy. In the course of that period, he was instrumental in establishing the voluntary marine reserve around the island in 1973 and undertook much of the work that now underpins the management of the statutory Marine Nature Reserve. Most recently, in 2006, he has re-surveyed some of the shores studied by Leslie and Clare Harvey in the late 1940s and early '50s and has contributed to the No-Take Zone monitoring. Keith Hiscock is a past Chairman of the Lundy Field Society and now an Honorary Vice-President. He is currently Executive Secretary of the Marine Biological Association at Plymouth and Programme Director of the Marine Life Information Network (*MarLIN*) there.

ROBERT IRVING

Robert first went to Lundy in 1983 when he was appointed the Nature Conservancy Council's Marine Liaison Officer for Lundy. During two consecutive summers spent on the island, his job was to facilitate the establishment of the country's first statutory Marine Nature Reserve, which came into being in 1986. Since that time, he has continued his interest in the island's marine matters. He has participated in various intertidal and subtidal monitoring studies around the island; led a number of conservation breaks for divers; undertaken an environmental impact assessment prior to the construction of the new jetty; and produced the short video film about the Marine Nature Reserve which is shown on board the M.S. *Oldenburg* when she sails to Lundy. Robert has served on the Committee of the Lundy Field Society since 1986, and co-edited the Lundy Field Society's 50th Anniversary book *Island Studies* published in 1997. Since 1990 he has been Secretary of the Lundy Marine Nature Reserve Advisory Group. He works as a consultant for his own marine environmental consultancy firm called Sea-Scope, based in N.E. Devon.

DR ROGER KEY

After gaining a BSc in Zoology at Nottingham, Roger went on to do a PhD in estuarine invertebrates and wading bird feeding ecology at Hull University. His first job was as Development Officer and Phase 1 Botanical Survey Officer for the Herefordshire & Radnorshire Nature Trust. He then joined Natural England (previously called English Nature), where he is now Senior Invertebrate Ecologist, dealing with the conservation of a wide range of insects and their habitats. He is also involved with the media, having been a natural history presenter of BBC2's *Countryside Hour* and *Langley Country*, and contributor to *The Natural History Programme* on Radio 4. Married to Rosy, in real life he is a keen gardener, cook, wildlife photographer and world travel fanatic.

ROSY KEY

Rosy is the Local Nature Reserves Officer at Natural England (previously called English Nature), promoting them via a website and the BBC *Springwatch* and *Autumnwatch* programmes. Her former post was deputy manager of the lottery-funded project 'Tomorrow's Heathland Heritage', restoring heathland over the U.K. Earlier jobs have been budget manager and exec officer, variously on conservation monitoring and site safeguard. She was formerly with JNCC doing international work on CITES and with the IUCN and earlier with the Nature Conservancy Council, over the years working on species advice, policy & planning and an entomological bibliography. Originally from Wales with a degree in Zoology from Hull University, she has always been a keen horsewoman, naturalist and hill walker and does lots of foreign travel.

TONY PARSONS

Tony Parsons is a veterinary surgeon, retired from a large Westcountry practice. His main interests are in ornithology and entomology, particularly involving migration of birds and insects and studies of parasitic Hymenoptera. He has been a bird ringer for 40 years, has taken part in a number of expeditions in Europe and West Africa and runs two bird ringing stations on the SSSI which he owns in south Somerset and on the island of Steep Holm where he is chairman of the trust which owns the island. Tony has been visiting Lundy for 50 years.

HENRIETTA QUINNELL BA FSA MIFA

Henrietta Quinnell is an expert on the prehistoric and Roman period archaeology, especially the ceramics, of South West Britain. From 1970 until 1999 she was lecturer in archaeology in the University of Exeter's Department of Extra-Mural Studies (later Life Long Learning) with responsibility for adult education courses throughout Devon and Cornwall. She has excavated widely in the region and published numerous papers in county archaeological journals. Her report on the excavation of Trethurgy Round - *Excavations at Trethurgy Round, St Austell: Community and Status in Roman and Post-Roman Cornwall* - published by Cornwall County Council in 2004 is the seminal work on Roman Cornwall. Since taking early retirement in 1999 she has been working as a consultant on prehistoric ceramics. She is currently President of the Cornwall Archaeological Society and a former President, now Vice-President, of the Devon Archaeological Society.

DR MYRTLE TERNSTROM

Myrtle Ternstrom (formerly Langham) first went to Lundy in 1952 and she has visited it regularly ever since. Her particular interest is in the island's history, and research is ongoing. She was part-author with Tony Langham of two books about the island, and since then has published another three, as well as being joint editor and publisher of F.W. Gade's Memoir, *My Life on Lundy*. Her enjoyment of Lundy arises from its tranquillity, the clear air, the wind, the sea and the birdsong; the friends she has made there, and the fact that it is an engrossing and yet encompassable subject of study. In 1999 Lundy's development, with consideration of comparable small islands, was the subject of her doctoral thesis.

JOIN THE

LUNDY FIELD SOCIETY

for the study and conservation of a unique island

Seals

Prehistoric hut circles

Sea birds

No-Take Zone

Marine Nature Reserve

Pond species

Industrial archaeology

Shipwrecks

Early Christian burials

Architecture

Lundy Cabbage

Geology

Medieval military architecture

T HE LUNDY FIELD SOCIETY is a charity for the study of the history, natural history and archaeology of Lundy, and the conservation of its wildlife and antiquities.

Whether you have just discovered Lundy or have known it for years, you will be welcomed as a member, and you will be making an important contribution to the study and conservation of the island through your membership.

For many of its members, the Lundy Field Society is an informal 'Lundy Fan Club' - a way of keeping in touch with the Island and with people who share the same interest and are happy to share their knowledge. The Society has a annual meeting, an Annual Report, a Journal and an annual Newsletter.

For more information about the Lundy Field Society, visit *www.lundy.org.uk*, and download a membership leaflet from *www.lundy.org.uk/lfs/join.html*.